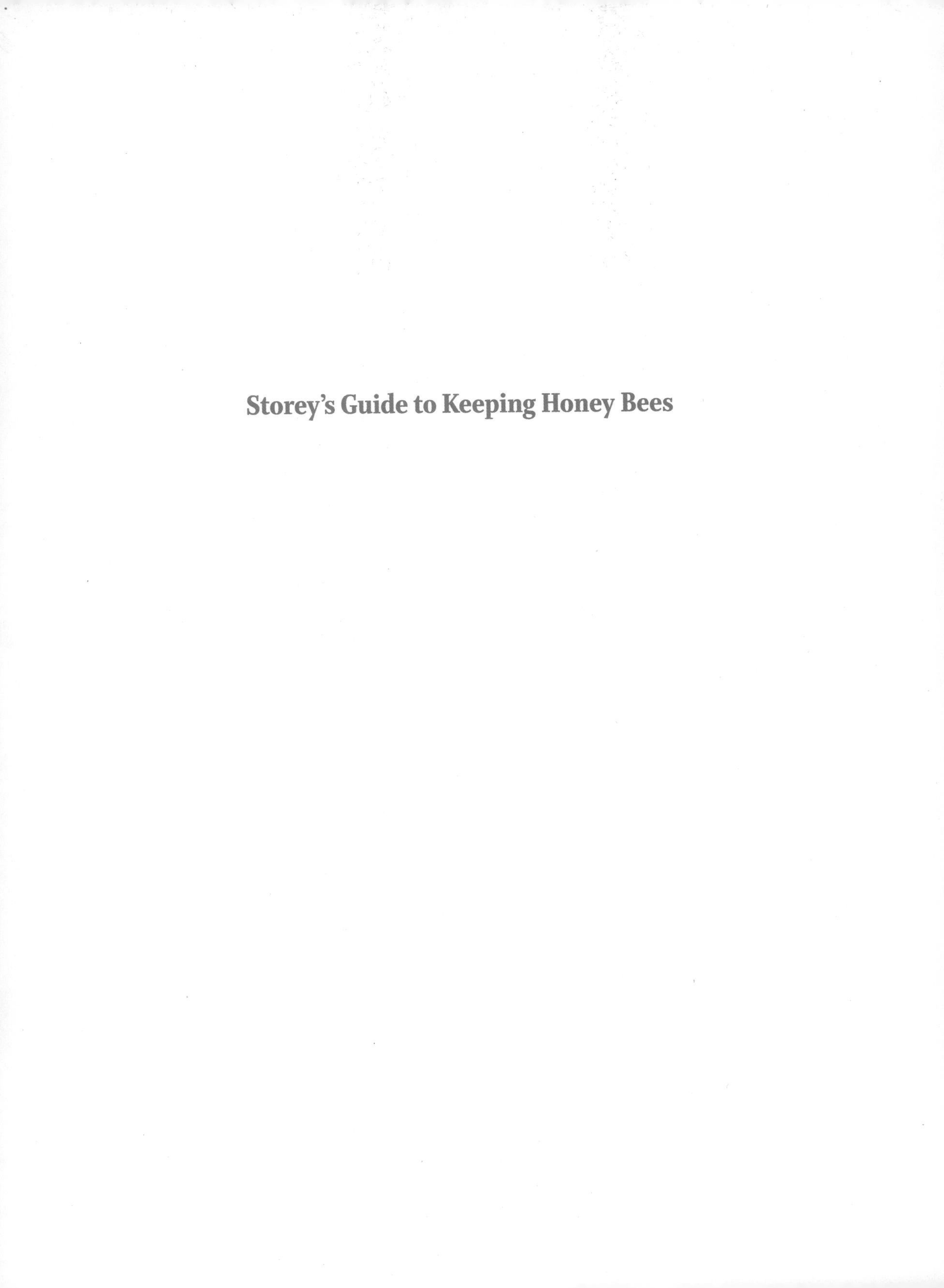

Storey's Guide to Keeping Honey Bees

STOREY'S GUIDE TO

KEEPING HONEY BEES

SECOND EDITION

HONEY PRODUCTION • POLLINATION • BEE HEALTH

Malcolm T. Sanford & Richard E. Bonney

Storey Publishing

The mission of Storey Publishing is to serve our customers by publishing practical information that encourages personal independence in harmony with the environment.

EDITED BY Deborah Burns and Sarah Guare
ART DIRECTION AND BOOK DESIGN BY Jeff Stiefel
TEXT PRODUCTION BY Liseann Karandisecky
INDEXED BY Nancy D. Wood

COVER PHOTOGRAPHY BY © Ed Reschke/Getty Images, front; © Monty Rakusen/Getty Images, back; © Riorita/Getty Images, spine
INTERIOR PHOTOGRAPHY BY © Mars Vilaubi, ii, 26, 34, 39, 42, 47, 49, 66, 68, 73–75, 76 bottom, 77 middle & bottom, 79 top & bottom, 80, 83, 86–94, 96, 97, 110, 120, 126, 127, 130, 135, 137, 140, 141, 146, 147, 158, 179, 190–191 and Mars Vilaubi, 52, 69, 70, 71, 76 top & middle, 77 top, 79 middle & left.
ADDITIONAL INTERIOR PHOTOGRAPHY CREDITS appear on page 204.
ILLUSTRATIONS BY © Elara Tanguy, based on original artwork by Elayne Sears, except 18 and 140 by Ilona Sherratt

The foundation for this book is two previous works by Richard E. Bonney, *Hive Management* (1990) and *Beekeeping: A Practical Guide* (1993).

Storey books are available for special premium and promotional uses and for customized editions. For further information, please call 800-793-9396.

Storey Publishing
210 MASS MoCA Way
North Adams, MA 01247
storey.com

Printed in China by R.R. Donnelley
10 9 8 7 6 5 4 3 2 1

Library of Congress Cataloging-in-Publication Data on file

I dedicate this book to my father, Malcolm Elam Sanford, who instilled in me the value of the written word. I became the published author he longed to be.

The book is also dedicated to the honey bee. This social insect gave me both the platform and the training ground to distill my thoughts into as few words as possible, while clearly communicating complex issues to a wide audience made up of scientists and laypersons alike.

— MTS

CONTENTS

ACKNOWLEDGMENTS

It took the assistance of a great many people to write this book. These include the scientists and curious laypersons who provided insight into honey bee biology over the last two centuries, as well as current associates in both lay and professional groups, who continue to share their knowledge and experiences with me. Thanks to the late Dick Bonney for creating the basic building blocks of the work, and to my editor Deborah Burns for her encouragement and assistance.

I especially want to express my gratitude to Dr. H. Shimanuki, friend and colleague, now retired as research leader of the Honey Bee Laboratory in Beltsville, Maryland. He reviewed the material in this work, contributing to clarity in his careful and insightful way, as was his custom when we collaborated throughout our professional careers. I would also like to thank Dr. Susan Drake, faculty member in the Family Medicine Residency at Tallahassee Memorial Hospital, Tallahassee, Florida, for her review of the section on bee stings and reactions.

All errors and omissions, of course, remain mine.

This volume provides a wider perspective than most of its kind through a sprinkling of new and experienced beekeepers' points of view based on different geographic locations, revealing yet again that "all beekeeping is local." These comments were contributed by current subscribers to my *Apis* electronic newsletter, in continuous publication for more than two decades (transcending my active career as Cooperative Extension apiculturist at two major universities). These unique, authentic voices cajole, persuade, empathize, and generally encourage all who would take up one of humanity's most challenging callings, culturing honey bees: Laurel Beardsley, Florida; Mark Beardsley, Georgia; Ed Carthell, Washington; Sharon A. Christ, West Virginia; Lynn Davignon, Rhode Island; H. E. Garz, Washington; Debbie Gilmore, Nevada; Dave Hamilton, Nebraska; Lawrence E. Hope, California; Jeffery Maddox, Missouri; Jeanette Momot, Ontario; Peter Smith, United Kingdom; Bill Starrett, Ohio; Paul van Westendorp, British Columbia; and Elise Wheeler, Massachusetts.

PREFACE

The genesis of this work was two volumes originally written by Richard Bonney: *Hive Management* in 1990, and *Beekeeping: A Practical Guide* in 1994. Dick owned and operated Charlemont Apiaries in Charlemont, Massachusetts, and later taught beekeeping at the University of Massachusetts in Amherst. He also served as a state apiary inspector and so had practical experience, as well as academic training, in managing honey bees. This is the perfect mix for writing about the beekeeping craft. It is indeed unfortunate that Dick is no longer with us to continue to act as a mentor to beekeepers.

As for my background, I managed honey bees at the University of Georgia research apiary, worked for a commercial queen breeder for a time, and received extensive academic training, serving as Extension beekeeping specialist at both The Ohio State University (1978–1981) and the University of Florida (1981–2001). I have published articles in U.S. and international beekeeping journals, traveled widely as an apicultural consultant, and presented papers at several international beekeeping congresses. It is an honor to be selected to carry on the work of Dick Bonney by updating his previous works in this now second edition of *Storey's Guide to Keeping Honey Bees*.

Beekeeping has changed a great deal since the publication of Dick's books. In addition, he wrote principally about beekeeping in the temperate portion of the United States. This reflected his considerable beekeeping experience in the Northeast, corresponding to U.S. Department of Agriculture (USDA) Plant Hardiness Zone 5, characterized by an average annual low temperature range of –20° to –10°F (–28 to –23°C). The advice in this volume covers a wider set of conditions as found in Zones 6 through 11. We also include stories, insights, and tips from beekeepers in different regions, to help readers find information relevant to their own situation. And finally, by necessity, the book looks at the craft on a larger, more global scale to reflect the realities of beekeeping in the twenty-first century.

The goal of this publication is to retain, as much as possible, Dick's style and content, while providing updated information apiculturists must know to be successful in the current beekeeping environment. The audience is the same as it was for Dick Bonney's work: the novice beekeeper. As such, this work focuses on beginning concepts and provides an introductory vocabulary of the craft. No book could hope to cover all the information that an apiculturist requires to manage honey bees effectively. It is important to keep in mind that although the honey bee has been intensively studied for centuries, there is much that scientists and beekeepers are still learning about what is perhaps one of nature's most complex creatures.

1

BEGINNING BEEKEEPING

Most new beekeepers come into this exciting endeavor with high hopes, and this approach colors their whole attitude. The craft appears to be casual and looks like fun, so why not try it out? Although enjoyable, beekeeping can be a disappointing failure. It requires preparation and commitment. It demands knowledge of the natural world.

The beekeeper also benefits from a certain amount of interaction with others in the apicultural community. Like the bees in their colony working together to survive, no individual human can succeed alone when it comes to caring for this social insect.

Eight Basic Tips for Getting Started

Beekeeping may appear daunting to the novice, with an overabundant supply of advice on how to ensure success. You cannot know all the pitfalls in advance, but you may find yourself loath to begin and suffering from a case of "analysis paralysis," trying to know it all before getting started. Once a moderate amount of information has been digested, however, it's best simply to plunge in. The following tips are suggested to ease the way.

1. **Start with new equipment** of standard (Langstroth) design and dimensions. Used and homemade equipment has the potential to create problems that the novice is not prepared to recognize or handle.
2. **Do not experiment during your first year or two.** Learn and use basic methods. Master them. This will provide a basis for comparison if you choose to experiment in future years.
3. **Before buying a so-called beginner's outfit,** know how each piece of equipment is used and be sure it is needed.
4. **It is best to begin with two bee colonies,** not one. The reason for this is the real possibility that as a novice you will lose a colony in the first season. Having one in reserve will provide a cushion against a complete disaster that might bring your beekeeping activities to a premature end. Two colonies will also be useful for comparative purposes.
5. **Start with Italian bees.** They are the standard in the United States and most commonly available. Those acquired from a competent producer should be gentle and easy to handle. In future years, you can experiment with other races or strains to form comparisons.

Starting with two colonies allows you to compare and contrast.

6. **Start with a package of bees** or a nucleus colony (nuc) rather than an established fully populated one. Establishing a nuc or package will help gain confidence. Again, if at all possible, start with two colonies to give a further basis of comparison. The shortcomings of one should be easier to detect, and a second unit can provide an invaluable resource — bees and brood to keep your beekeeping operation alive in case of emergency.
7. **Start early in the season,** but not too early. Seek guidance from local beekeepers and bee inspectors about timing in your specific area. Remember that all beekeeping is local. Advice about managing honey bees from those in other geographic areas, even if they are successful, is fraught with risk.
8. **Recognize that your colonies will not produce a surplus of honey** the first year, especially those developed from **package bees**. The first year is a learning time for the beekeeper and a building time for the new honey bee colonies.

Dimensions of Beekeeping

Beekeeping is multifaceted and involves much more than placing a hive in a backyard, visiting it a couple of times a year, and reaping the rewards. In fact, it can require a great deal in terms of physical effort, time, and money. The beginner can easily be overwhelmed by having to juggle all the balls required to manage a honey bee colony effectively. The best course of action is to have a plan, start small, and gradually progress in discrete increments, all the time carefully monitoring your goals and modifying them as conditions warrant.

Of all agricultural enterprises, beekeeping may be one of the most difficult to manage. Most livestock are at least under a measure of control. Cattle, horses, and hogs can be tethered, fenced, or otherwise confined. Crops do not move, although this can also be a disadvantage when it comes to their management. It is relatively easy to measure the feed, money, time, and effort a manager might put into most agricultural operations against what is produced. Honey bees, in contrast, are free-flying insects, and a good proportion of a colony's individuals may be foraging within a two-mile radius of the hive. Much of the time the beekeeper has little knowledge of where the bees are, or what they might be doing.

The beekeeping craft continues to undergo rapid changes in our modern, fast-paced world. The human-facilitated movement of biological material around the globe has reached epidemic proportions, and the honey bee has not been spared the unanticipated consequences, including the introduction of alien pests and diseases. Challenges, therefore, can easily become global in nature and often take beekeepers and researchers by surprise. These can quickly add other dimensions to an activity that, in the past, was often considered traditional and unchanging.

Beekeeping Objectives

Traditionally, beekeepers have been pigeonholed into categories. Most common are the terms "hobbyist," "sideliner," and "commercial," generally based on the number of colonies being managed. Unfortunately, these labels often do not reflect the actual situation; they are nothing more than reference points along a continuum of the beekeeping experience.

Another problem with these generalizations is that in the modern political climate, words mean a lot. The crafting or framing of messages is now an art. Legislators who are approached to assist beekeepers might be confused about designations such as "hobbyist," as well as other terms defining apiculturists like "sustainable," "organic," "small-scale," and "artisanal." It is probably best to label anyone in the craft simply a beekeeper — a person who cares for and about honey bees — and let it go at that.

A better way to classify beekeepers is by objective. Most are interested in producing **honey**. Others might principally become pollination managers because they grow their own plants or are near growers who need the service. If honey is your objective, is the goal of marketing it not far in the future? The same can be asked about pollination services.

The craft may indeed be a true hobby or pastime. Many significant advances in beekeeping were not developed by commercial

BEE FEVER

THE UPSIDE

The upside to beekeeping usually outweighs the downside. Why else would so many have persisted in carrying on the activity for thousands of years? Go to any gathering of beekeepers and listen to them talk (even about their challenges) with enthusiasm and pride. Go into a beeyard on a pleasant day and sit there, immersed in the calm serenity of the scene. Watch the bees coming and going, sometimes even indulging in what is called "play time." If beekeeping is truly in the blood, it is impossible to resist the allure of the craft. For want of a better term, some call the passion that arises in some novices "bee fever."

THE DOWNSIDE

Beekeeping is not all pleasure. As with most things in life, it has a downside. Some beekeepers have a mentor or partner, but most are alone when they begin. They are out there by themselves, in the heat, sticky to the elbows, bees buzzing about, and with a veil in place so it is impossible to scratch or blow a nose or drink water.

The bees may become defensive for no particular reason. In this state, they invariably find ways to get under the veil or up the sleeves and pants legs, approaching places that are not polite to discuss when company arrives. There may be no choice on occasion but to leave the field of battle in search of a quiet place to rest and recuperate before again entering the fray.

It is not wise to discount the heat, weight of protective equipment, or discomfort when working in a beeyard. There are times when it becomes obvious that things are not under control, when you ask how you got into such a position, with hive parts strewn about, bees everywhere, sweat stinging the eyes, and a buzzing bee under the veil. Fortunately, these times are the exception rather than the rule and usually quickly forgotten, as bee fever takes hold again.

beekeepers, in fact, but by visionaries who also studied philosophy and religion.

This book cannot speak to all beekeepers, and no volume is worth much that does not have an audience in mind. In this case, the audience is the novice who wishes to explore the craft, with the possibility that his or her beekeeping activities might expand far beyond current expectations. Typically, that person starts out as a honey producer; thus, the focus of this volume is producing and processing the sweet reward that first caused humans to hunt, and later to keep, honey bees.

The Beekeeper's Commitment

Too many novice beekeepers do not recognize the level of commitment they must have to successfully manage honey bees. For every beekeeper who succeeds, there are probably two or three who do not. You may have been in a classroom or training program where the instructor begins by saying, "Look at the person on your right; now look at the person on your left. One of you won't be here next year or next week or next month." Beekeeping is like that.

Not all beekeepers are truly bee-keepers. Some are bee-havers. The latter develop an initial enthusiasm, acquire some bees, and work with them for a while, eventually losing interest. They never develop any real knowledge or skill. In the end, they have bees, perhaps a colony in the backyard, but are not really keeping bees. Present-day beekeeping challenges continue to winnow out many of these less-than-committed beekeepers. This actually strengthens the beekeeping community in the

BEEKEEPER'S STORY

MY OWN APIARY has doubled in size since my beginning, to the grand ol' number of two colonies. Although I lost the original queen this spring from each of those hives, I can boast that at least I carried my bees through winter and into spring. I designed and built my own lightweight but sturdy hive stands from PVC pipe, which was slightly criticized and doubted by the male species of beekeepers who saw my setup. While I admit that I was fortunate not to have any foot-deep snows piled atop my hive, they are both still standing.

My hives were inspected in May, and no disease or pests were found. I have applied no commercial chemicals but relied on my "motherly instincts" to keep my bees healthy.

A current favorite pastime of mine involves following my bees around with a camera, which also tells me which flowers are being worked and when. My bees are currently working late sumac.

One of my most interesting observations is the way the bees collect pollen from the wild daisy patches that I mow around. They attach to the center of the flower with their mouths and drag their bodies around the flower in a circular motion collecting pollen on their legs. It almost appears as if they are deformed.

Sharon A. Christ, West Virginia

long run because the members who remain are well informed and truly passionate.

A successful beekeeper learns about honey bees, comes to understand them, and works them on a regular basis, enjoying the process. In the final analysis, the best beekeeper is able to "think like a honey bee." Managing honey bees, therefore, is as much "art" as it is science.

The Beekeeping Community

Beekeeping is a dynamic endeavor. Problems arise, solutions emerge, research is undertaken, and new knowledge continually comes to the fore. It is difficult to keep bees effectively without connecting to other beekeepers and acquiring new skills demanded by the craft. Much good advice comes from government agencies; universities; and local, state, national, and international beekeepers' associations. All of these organizations are an important part of the overall beekeeping picture. At a minimum, the beginning beekeeper should join a local beekeeping club or association and regularly read a beekeeping magazine.

Some individuals take up beekeeping because they want honey bees for pollination, whether for a small home garden or a commercial growing operation. They may obtain one or several colonies, set the bees up in a far corner of the property, and forget about them, assuming they will take care of themselves. Unfortunately, the facts of honey bee life include disease, drought, harsh winters, and predators; all of these can cause a colony to weaken and perish. It happens regularly in nature—too regularly, perhaps. Thus, a colony of honey bees often has a tenuous grip on life, especially in more northerly regions. People contemplating installing bee colonies strictly for pollination should seriously consider finding a committed beekeeper to look after the hives they've installed.

Time Commitment

The time devoted to keeping honey bees does not have to be great, but as with many activities, you get back proportionally what you put into it. The time commitment also varies with the beekeeper's goals. The largest time investment will be in the learning phase as you begin the craft: reading and attending bee schools, workshops, and meetings of local and national associations. Plan on a learning curve that will be steep the first year or so, and continue to expect challenges throughout your beekeeping career.

Many individuals undertake beekeeping with minimal preparation, believing they can simply dive in and pick up the requisite knowledge. More feasible in the past, this has become an increasingly less successful strategy as honey bees have become much more vulnerable to the exotic pests and diseases that continue to plague them, and by extension, the beekeeper.

Visiting the Bees

Plan on visiting the bees an average of every two weeks during the active season, perhaps more often as the new season is getting underway, and less often in the inactive part of the year. Individual examinations per hive can be quite brief, depending on the season and reason for being at the hive. Some may last a minute or two; others involving a specific task might take 20 to 30 minutes per hive at the longest. Most inspections that involve opening the hive are a substantial disruption to colony life, and so should serve an important purpose. Have a goal in mind before disturbing the insects.

Bee colonies should never be totally ignored. Some endeavors can be picked up or put down at will. Not beekeeping. Chores not done at the proper time cannot usually be done effectively later, if at all.

Record Keeping

Most beekeepers don't keep adequate records. It's that simple. Comprehensive long-range bee management and financial record-keeping systems are essential to maximize efficiency and maintain profitability in any apicultural enterprise.

A visit to any beeyard usually reveals colonies marked by sticks, stones, or other readily available materials placed on each colony's cover. The specific arrangement of these materials may indicate everything from status of the queen to whether a colony requires food.

Visit the bees every two weeks or so during spring and summer, always with a purpose and a goal in mind.

Two problems arise with this kind of record keeping. It is short-range; once the status of the colony changes and the materials are rearranged, the previous information is lost. The system is also unique to each operator, rendering it nontransferable and nontranslatable to others. Part-time employees or bee inspectors are unlikely to know the meaning of such records. Marking queens is also essential. It permits the beekeeper to judge productivity of individuals purchased from different breeders and to determine a queen's age and whether supersedure has occurred.

Perhaps the biggest problem with record keeping is deciding what information to save. The sheer volume of potential data that can be collected is staggering. The beekeeper must, therefore, carefully choose what is most important for each particular enterprise.

Financial Considerations

Most beekeepers view the craft as a hobby at first, but often quickly begin to ratchet up their activities. This can cost a good deal of money. The key is to adjust your goals to the potential costs involved. Beekeeping is inherently a risky enterprise, and far too many variables and problems exist for any beekeeper ever to be completely in control. As with any farm activity, it is highly dependent on natural phenomena such as weather, rainfall, and the state of local vegetation. As mentioned, however, it differs from other agricultural enterprises because the beekeeper simply lacks much of the control inherent in other farming activities.

Some beekeepers get into trouble if they base their expenditures on dreams rather than reality. Expect it to cost more money than you think in the beginning, and that it will take more time than estimated to recoup your investment, no matter the future goal. For now, assume an expenditure of several hundred dollars, depending on sources and quality, to set up a complete hive with a couple of honey supers (special hive boxes for collecting honey), with a queen and bees included. At this point in time, it takes a beginning capital of at least $300 to begin a single colony and purchase related equipment. As mentioned, it is best to begin with two hives, which adds perhaps another $200.

Accounting

One thing many beekeepers do not emphasize at the beginning of their career is financial record keeping, but it is essential as the enterprise grows. Modern business success is predicated on the use of balance sheets and income statements as part of a long-range strategy. Commercial beekeeping is no exception.

One Step at a Time

Go into beekeeping strictly as an enjoyable hobby. Expect it to cost money initially. It will be all outgo for a while. If, after a couple of years, you have surplus honey to sell, that's wonderful. In time, you may even have enough hives to move into crop pollination in a small way. Eventually, if all is going well, perhaps you can turn this hobby into a sideline business, but don't base your future on it now. Wait until you see what it's all about.

Stings

Honey bees sting, and being stung is a fact of beekeeping life. Stinging is defensive behavior. This cannot be said too many times as most people have been conditioned to think of honey bees as aggressive. In fact, they are simply reactive and defensive. Here are some essential points about stings.

Most often stinging is prompted by behavior the insect itself deems aggressive, especially if the nest appears to be threatened. Swatting at bees, approaching too close to a nest entrance, and vibrating the hive or nest are examples of especially threatening human behavior. If a colony has been recently disturbed, stinging behavior will most likely increase for a period of time and may not subside until the next day. Some people are more prone to being stung than others. This may be due to body chemistry or to fragrance (perfume, deodorant, hair spray) that is attractive or offensive to bees.

Bees are recruited to become defensive by what is called an **alarm pheromone** (a chemical signal, usually an odor). This principal substance has been intensively studied and is known as isopentyl acetate, which smells like bananas to humans. A suite of other pheromones may be involved as well. The reason for using smoke on colonies is to mask the odor of these substances.

Beekeepers get used to stings fairly quickly and, in some cases, may not suffer much pain at all, depending on the site. Stings on the head, face, or fingertips are usually the most painful. A bee sting, however, is always potentially serious. The severity and duration of a reaction can vary from one person to another and from one time to another. In addition, one's own reaction to a bee sting may differ between occurrences.

Most people experience a local, nonserious reaction to bee venom. Depending on the location and number of bee stings received, however, as well as the ever-present possibility of a severe allergic reaction to bee venom, a sting can precipitate a life-threatening situation.

"I GOT STUNG!"

The honey bee cannot withdraw her barbed stinger once it has penetrated the human skin. The only means of escape is to tear away part of her **abdomen**, leaving behind the stinger with its venom sac attached. The muscles of the sting apparatus continue to pulse after the bee has flown away, driving the stinger deeper into the skin and injecting more venom. For this reason, the sting apparatus should be scraped out of the skin as soon as possible after a sting is received. The offending bee may continue to harass, but it can no longer sting and will soon die.

When someone gets stung:

- Notify a companion in case assistance becomes necessary.
- Immediately scrape the sting apparatus out of the skin.
- Wash the site and/or apply rubbing alcohol to reduce pain.
- Apply ice to help reduce swelling.

There are a great many remedies for stings suggested in the informal literature, including tobacco juice, toothpaste, meat tenderizer, and human saliva. The bottom line, however, is that the venom has been injected into the skin and there is no way to remove it.

The only effective first aid for extreme systemic reactions to honey bee venom is injecting adrenaline into the human body. Commercial "sting kits" containing epinephrine (adrenaline) can be found at any drugstore and usually require a prescription by a physician. No beekeeper should be out in the beeyard without such a kit, which often includes both antihistamine tablets and an epinephrine syringe for emergencies. This goes double for anyone hosting human guests in their beeyard.

When visitors are present, at a minimum, develop a plan to assess their history of being stung. If symptoms of a systemic reaction appear, stabilize victims with injectable epinephrine found in a sting kit, and immediately transfer them to an emergency room.

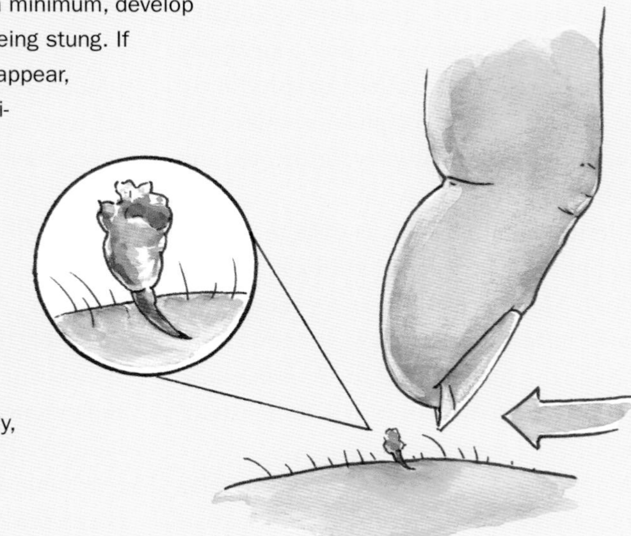

SCRAPE OFF THE STING APPARATUS
The most important thing to do when you are stung is to remove the stinger immediately, preferably by scraping it off.

Children often suffer more from stings than adults do. This may be psychological or because of lower body weight. Keep in mind, however, that young folks are the foundation of the future beekeeping community. Encourage them to look at honey bees from the beekeeper's perspective — as beneficial insects rather than as creatures out looking for trouble. Don't make too much of a bee sting; put something on the site and give plenty of sympathy to the child, but then continue as if nothing has happened.

Reactions to Stings

Two kinds of reactions are usually associated with insect stings: local and systemic.

A **local reaction** is generally characterized by pain, swelling, redness, itching, and a wheal surrounding the sting site. This is the reaction of the vast majority of people, and this group is considered to be at little risk of death. Many in the general population, however, believe that because they "swell up," they are at risk of losing their life when stung by bees.

Ironically, it may in fact be the reverse. Those far more at risk may show no effects from stings at all until they suffer a **systemic reaction**, sometimes called **anaphylactic shock**. This is most dangerous when the mouth, throat, and/or airways are affected and respiration is interfered with.

Allergists and physicians disagree about how to approach anaphylactic shock (or anaphylaxis). Many prefer to err on the side of extreme caution, treating most systemic reactions as life threatening. But this bias has little basis, according to Dr. Howard S. Rubenstein, writing in *The Lancet*:

"Many of the large number of people who are stung each year by bees experience frightening systemic reactions," Dr. Rubenstein writes, "but the vast majority of such reactions are not life threatening. There is no evidence that the very few who die as a result of a bee sting come from the pool of those who once before sustained a systemic reaction. On the contrary, no reaction at all may be a more ominous predictor of a lethal outcome of a subsequent sting."

Death from bee stings comes about through a number of mechanisms, according to Dr. Rubenstein. Perhaps most important is the effect of **atherosclerosis** (buildup of deposits in the arteries) and unrecognized cardiovascular disease. External factors such as the temperature and the site of the sting also affect mortality.

"A patient who suddenly develops hives, shortness of breath (sometimes with bronchospasm), and giddiness or **syncope** (fainting) sometimes with **hypotension** (drop in blood pressure) is terrified, as are those about him," writes Dr. Rubenstein. "The patient may think he is going to die, as may his family or physician. What people need to know, therefore, is that the vast majority of patients, particularly if aged under 25, will quickly recover."

In conclusion, the dilemma facing both physician and beekeeper is deciding whether a frightening yet self-limited response needs immediate medical treatment.

Tolerance and Treatment of Stings

Beekeeper tolerance of stings increases with exposure. After a few stings, many realize they are no big thing and begin to ignore them when they occur. One of the authors once noticed the grayness of one experienced beekeeper's hand at a busy apiary. On closer examination, it became apparent that his hand was literally covered with dried-up stingers that he had not bothered to remove.

STING REACTIONS

COMMON NORMAL REACTION
Whitened wheal, central red spot, localized swelling and pain, subsiding in minutes to hours. Large, local reactions may also occur, subsiding in a few days.

RARE ABNORMAL REACTION
Systemic reactions sometimes show little localized swelling, but can affect organs distant from the sting site. These may result in hives, trouble breathing because of airway swelling, and/or drop in blood pressure.

MULTIPLE STINGS
Any person, regardless of sensitivity to bee venom, receiving an enormous number of stings (**mass envenomation**) might be susceptible to renal failure or other severe symptoms simply because the body was challenged by a great quantity of toxin. In places where Africanized honey bees are established, physicians should retrain themselves for the latter possibility, where the first-aid technique is renal dialysis, not injectable epinephrine.

Over time, the human body also becomes accustomed to the venom. The results of stings in terms of pain and swelling are, therefore, often reduced. The reaction to stings may shift over the beekeeping season, being more extreme at the beginning when beekeepers first begin to visit colonies.

For some, however, tolerance to stings can take a different course. These individuals react more with each successive sting and can develop a true allergy requiring medical assistance. This can happen to beekeepers as well as non-beekeepers. It has been observed most closely in beekeepers' families, especially in those members who are not active in the craft. Apparently, the repeated breathing of particles of dried venom or other bee-related materials that adhere to beekeeper's clothing can sensitize other family members. Be sure to wash beekeeping clothing frequently, and keep clothing and other beekeeping paraphernalia where others are not exposed to it.

The best advice for beginners is to always protect your face with a bee veil. This alone will prevent most stings, especially the most painful: those to the lips, nose, eyes, or ears. That said, get used to regularly taking a sting or two to your hands, to build up a healthy tolerance to bee venom. Attempting to avoid *all* stings could predispose you to an eventual full-blown allergy.

Stings by Africanized Bees

The African honey bee was brought to Brazil in the 1950s, and its crossing with the already-established European bee resulted in a hybrid known as the **Africanized honey bee**. This insect gradually infiltrated the southwestern United States and Florida. Although disease-tolerant and a good brood-producer, the Africanized honey bee has a reputation for excessive swarming and defensiveness. Since the 1970s, in fact, sensationalized reports of its stinging behavior have bombarded the general public. Dr. Mark L. Winston, in his book *Killer Bees: The Africanized Honey Bee in the Americas* (1993), called it the "pop insect" of the 20th century.

Experienced beekeepers shrug off stings. Any person, regardless of sensitivity to bee venom, receiving an enormous number of stings (**mass envenomation**) might be susceptible to renal failure or other severe disorders when the body is overwhelmed by a great quantity of toxin.

In places where Africanized honey bees are established, physicians should prepare themselves to treat with renal dialysis rather than injectable epinephrine. The lethal dose of honey bee venom in humans is around 19 stings per 2.2 pounds of body weight or about 1,300 stings for a 150-pound person.

Incidence of allergic reaction to insect sting, according to Dr. Camazine (see the box below), probably occurs in less than 1 percent of the population, and only a small percentage of those with an allergy develop severe reactions. Even with the arrival of the Africanized honey bee and associated stinging incidents, Dr. Camazine concludes there is no reason to suspect that bee stings will become a significant health hazard. Certainly, there is more reason to be concerned in areas where Africanized honey bees are established, but still the odds of being attacked are extremely small. The Africanized honey bee continues to become more entrenched in North America, however, so it pays to know where populations are established. Nevertheless, fewer than 40 deaths from stinging incidents have been recorded in the United States where this bee is established since its introduction in 1990.

PUTTING STINGS INTO PERSPECTIVE

Dr. Scott Camazine, writing in the *Bulletin of the Entomological Society of America*, says that most people have a great fear of venomous animals. In the bigger picture, however, he says, insect stings are a minor health problem. About 40 deaths occur each year because of stinging insects, most in the order *Hymenoptera* (ants, bees, and wasps); honey bees may cause half. Allergic reactions to penicillin kill seven times as many people and lightning strikes kill twice as many. In contrast, the nation's largest killers are cardiovascular disease (100 people per hour) and automobile accidents (one person every 10 minutes). Ironically, Dr. Camazine says, one is at more risk of dying in an automobile accident on the way to the hospital to be treated for an allergic reaction than of dying from the sting that produced it.

Effect of the Sting on the Bee

If it is any consolation, when a honey bee stings, she gives up her life (see box on page 9). She does not die immediately: she will often continue to harass a perceived threat to her hive and to recruit others to help. The sting apparatus remains in the victim, however, torn from the bee's body, and she is irreparably damaged and dies.

Biologists have intensively investigated the source of this altruistic behavior over many years. It fascinates many people, and is one of the reasons they become interested in social insects, perhaps ultimately taking up beekeeping.

Legal Considerations

Among the risks of beekeeping is the possibility of legal repercussions. Because honey bees sting, they are often looked at askance, even with dread, by members of the general public. Therefore, if a beekeeper runs into legal trouble, few from the larger human population will offer aid and assistance. The threat of anti-beekeeping ordinances is very real in most communities. Most municipalities have rules about nuisances, which can quickly implicate beekeeping activities. Any adverse publicity will inevitably pressure beekeepers to give up the activity. This is especially true in any area where Africanized honey bees have become established. The best legal advice for the novice beekeeper is to become the best of neighbors and to maintain a low profile while keeping honey bees.

There are many examples of beekeepers working with local officials to ensure that beekeeping is not banned outright in municipalities. This cooperation has occurred in Denver, San Francisco, Vancouver, New York, and other cities, as well as in small towns. The beekeeping community's usual strategy is to support an ordinance that licenses beekeepers or regulates their behavior in some way, such as limiting the number of colonies in certain areas. An approach in Florida, where beekeepers voluntarily sign compliance agreements based on best management practices (BMPs), has met with some success.

Laws and Regulations

There is no federal regulation about beekeeping in the United States, aside from a 1922 law and amendments restricting the importation of honey bees and genetic material from other countries. This legislation, passed specifically to prevent the introduction of the **tracheal mite, *Acarapis woodi***, remains in force to this day. This means that no queens, adult bees, sperm, or eggs can legally be brought into the country from other parts of the world. In some instances, the regulations arising from the legislation have been revised to permit entrance of bees and limited introduction of sperm and/or eggs from other regions.

Many states have promulgated rules about beekeeping. Some are designed to regulate movement of honey bees between states, whether by packages in the mail or for commercial pollination services. Most bee inspection programs were enacted to help control a persistent and pernicious disease called **American foulbrood** (AFB). Over the years, the practice of bee inspection has ebbed and flowed, depending on financial resources. This activity usually is constituted under a state's agriculture department and plays an active role in evaluating honey bee health and well-being.

It is imperative that novices find out about the laws in their specific state. Some states require beekeepers to register and pay a fee for inspection services. A national association of apiary inspectors, consisting of representatives from states with active inspection services, meets at least once a year.

Food-safety regulations may also affect beekeepers planning to process, pack, and sell honey or other products. Food safety is usually part of the regulatory apparatus and administered by a state agriculture department and the U.S. Food and Drug Administration (FDA).

Regulations are likely to shift over time as circumstances change. Again, the best advice is to develop a relationship with the state agriculture department and the food safety inspectors it employs. Beekeepers' associations are invaluable in this activity. When important issues arise, groups of like-minded citizens have a stronger voice than do individuals.

Finding Resources

An entire book could be written about resources available to beekeepers. In particular, the advent of the personal computer and Internet has brought directly to the beekeeper's desk a

rich trove of information about honey bees and **apiculture**. Much of it, however, is either not suitable to the novice or can be downright misleading. The following tips are suggested to help navigate the maze created by a surfeit of advice, courtesy of "the information age." See also the resources in the back of this book.

Try to find an experienced, successful beekeeper who is willing to help. A good mentor is hard to find, but worth the effort once located. Remember, though, that beekeeping experience does not necessarily mean competence. Some beekeepers have only one experience, the same year repeated many times over. Others who are successful may not have the capacity and patience to train a new beekeeper. Ask a lot of questions and immerse yourself in information from other sources so as to compare it with what your mentor says.

Read. Read. Read. Volumes have been written about the craft. More has been written about bee biology and management than any one person could possibly read, so be as selective as possible. Realize that huge changes have occurred in beekeeping and in our environment, especially in the past 20 years, and most printed volumes have good content but must often be augmented with current information. Much of this is now available through online resources, including electronic newsletters, discussion groups, and forums.

Join local, state, and national beekeepers' associations now, even if you don't yet have bees. These groups encourage beginners and are a prime source of information about local beekeeping conditions.

Enroll in beekeeping classes held around the country, often sponsored by local and regional associations and/or the Cooperative Extension Service of the state land grant university. There is an Extension office in almost every county in the nation. Master Beekeeper programs are also found in many states (such as Florida, New York, and North Carolina). Similar programs are found through regional groups such as the Eastern Apicultural Society, the Heartland Apicultural Society, and the Western Apicultural Society.

Subscribe to national beekeeping magazines. Some have been in continuous print since the 1860s, and provide discounts to beekeepers who are members of local and state associations.

With all of these thoughts in mind, let's get started.

CITY ROOFTOP BEEHIVE: Many cities and towns in North America and around the world have developed provisions to allow beekeeping.

2

ORIGIN AND HISTORY OF BEEKEEPING

It is possible to keep honey bees without knowing much about the insect's biology or background. You will be a much more effective manager, however, if you understand the dynamics of the insect and its relationship to the colony, other organisms, and the environment. Beekeepers keep only one species of insect, the Western honey bee known as ***Apis mellifera***, meaning "honey bearer" or "honey bringer." There are tens of thousands of species of bees, however, and a lesser but no less important number of species of related wasps.

The general public tends to throw all stinging insects into a common pot called "bees." Although bees and wasps may look similar, they are not. Bees are vegetarians, consuming only sugary plant secretions, called **nectar** (carbohydrate) and floral **pollen** (protein). Wasps also imbibe nectar, but they take their protein from other organisms; they are carnivorous and often prey on other insects that might be human or bee pests. Both groups are beneficial insects and should be protected, although many in the general public don't realize this.

Honey Bee Evolution

Wasps and honey bees are closely related: in fact, honey bees are thought to be the descendants of flesh-eating wasps. The following steps were important in the evolution of present-day honey bee society, which continues to exhibit these behaviors:

Individual feeding. A single, primitive mass feeding of the young, as practiced by solitary bees, gave way to progressive feeding, providing the young with predigested food when it was needed.

Communal society. Gradually, a single reproductive **queen** and her infertile daughters began to live together as a **colony**. By virtue of the division of labor by castes and subcastes, that colony could then function as a long-lived **superorganism**.

Complex information network. Communication among individuals evolved into an extremely complex system, which included food sharing (**trophallaxis**), information exchange by chemical **pheromones**, and elaborate dances.

Established nests in trunks of trees or "gums" were some of the first managed colonies.

APIS (HONEY BEE) FAMILY TREE

Bees and wasps share the following classification:

Kingdom. Animalia

Phylum. Arthropoda: insects, spiders, crustaceans

Class. Hexapoda (six footed) or Insecta

Order. Hymenoptera (membrane-winged): bees, wasps, ants, sawflies, horntails

Suborder. Apocrita (referring to narrowing or constriction of the abdomen): bees, wasps, and ants

Bees diverge from wasps at the superfamily level:

Superfamily. Apoidea: bees. Wasps and ants are in the Vespoidea superfamily

Genus. *Apis*: honey bees

Species. *Mellifera*: Western honey bee

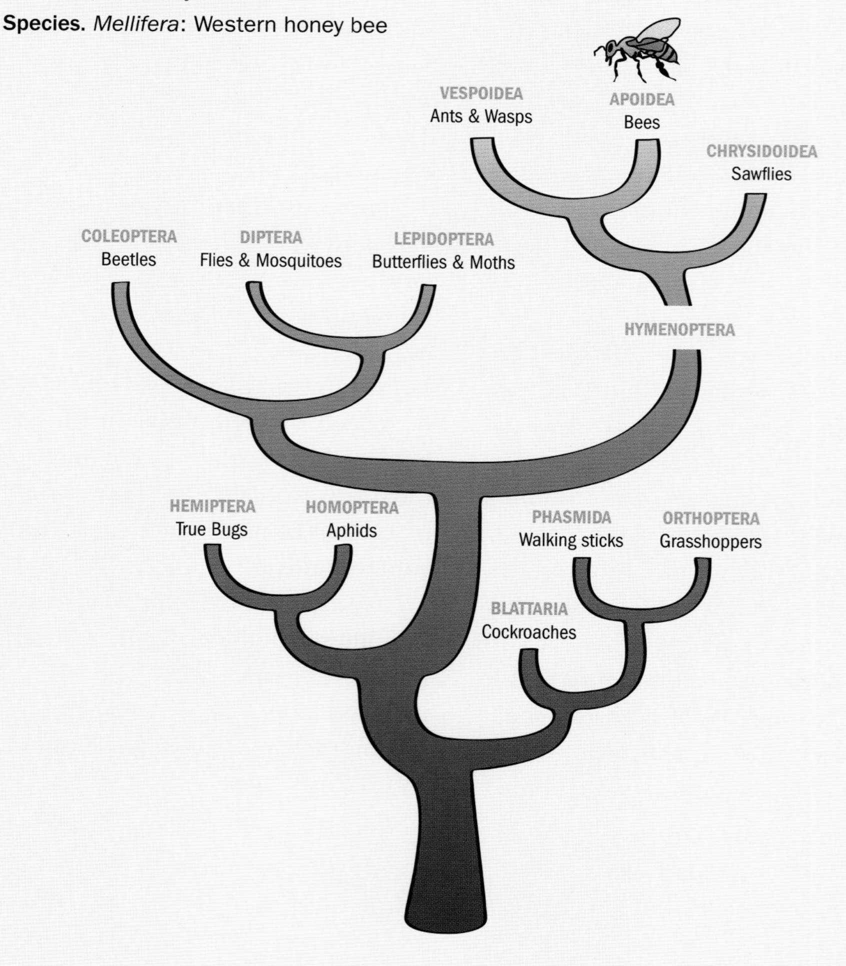

The western honey bee, *Apis mellifera*, is a single species that has migrated and been transported by humans around the world. Scientists once thought honey bees originated in Asia. New DNA evidence, however, suggests that the species arose in Africa and then migrated westward, through North Africa and the Iberian Peninsula into Europe, at the end of the last glacial period.

Honey Bee Ecotypes

Over the centuries, honey bees have sorted themselves into a number of subspecies or races of discrete populations based on local climatic conditions and local plants they are able to use as nectar and pollen sources. These subspecies are called ecotypes. Thirteen exist in Africa, six in the Middle East and Asia, and about nine in Europe. Those brought to the New World have been mostly European in origin and include:

- the German or black honey bee, *Apis mellifera mellifera*
- the Italian honey bee, *A. mellifera ligustica*
- the Carniolan honey bee, *A. mellifera carnica*
- the Caucasian honey bee, *A. mellifera caucasica*

Non-European introductions to the Americas include:

- the Egyptian honey bee, *A. mellifera lamarckii*
- the Syrian honey bee, *A. mellifera syriaca*
- the African honey bee, *A. mellifera scutellata*

All of these ecotypes can interbreed, which means that hybrids also can be found.

ALL BEES ARE NOT ALIKE

There is a great deal of published information describing differences among the honey bee races now found throughout the world. Italian bees (*Apis mellifera ligustica*) produce large amounts of brood year-round, are very quiet on the combs, and are somewhat more resistant to American foulbrood than others. Carniolans (*A. mellifera carnica*) rapidly adjust their brood rearing to the season and have a medium-length tongue. The short-tongued German or black bee (*A. mellifera mellifera*) is often defensive and susceptible to disease, while the Caucasian (*A. mellifera caucasica*), with the longest tongue, is prone to foulbrood infection and collects a great deal more propolis than the other races.

Throughout the history of beekeeping in the United States, the four major European races of introduced bees mentioned above have, as has the human population, lost their individual identity and disappeared into a great melting pot. A good deal of advice has been written about maximizing honey production, controlling swarming, and other management practices over the years. In the cacophony, however, the predominant race of bee being kept is often ignored. That's understandable because it is difficult to tease out specific groups that inhabit a particular colony.

Giving counsel or making management decisions that discount the fact that all bees are not alike can be counterproductive. It is not necessarily the ability to react to the status of individual hives, but rather managing the variability among colonies that better defines the true bee master.

The Superorganism

Honey bees are social insects. This means that although individuals are the basis for any colony, they cannot exist in isolation without help from other members. The colony's efficiency is what makes this insect so successful wherever it is found, and this is the basis for its management by humans. Beekeepers, therefore, do not manage single bees, but a colony of individuals, sometimes called a **superorganism**.

History of Beekeeping

Most beekeeping technologies were developed in the 1800s; the only technique considered a child of the 20th century is instrumental insemination.

The history of apiculture in some ways parallels that of science itself. From prehistory through the Dark and Middle Ages, little was known about honey bee biology. A rock painting dating from about 6,000 B.C. in Cueva de las Arañas, Spain (see page 22), is reproduced

BEEKEEPER'S STORY

A NUMBER OF YEARS AGO, my wife and I purchased a home on 15 acres in the hills of southwest Washington. Our main goal in having the property was to have a place for our horse, but over the years we have added many a critter. My wife also wanted to have an orchard so that we could have our own fruits.

At that time, I was working as the plant manager for a steel fabricator that hired a lot of immigrants. Most could not speak English but were hard workers who never missed a day. To help with communication, I ran a program that helped teach workers how to speak English. Each day, I would write ten English words on a blackboard and have them translated into Russian, a language all of the immigrants could speak. This program was in its fifth month when Valentine, one of my workers, had a question.

"He have farm, land?" he asked. "I keep bees, I keep bees your home," he said. I thought I understood him and asked what he needed. He said, "Land and orchard." We called over one of the other workers who spoke better English, and he explained that Valentine needed a place to keep his beehive. I told him it shouldn't be a problem, but that I had to ask my wife. Little did she know.

A week later Valentine moved in ten hives and started to maintain them and build new ones. He worked his bees two to three times a week during the milder seasons and at least once a week in winter. Over the next year, about every two or three months, we would come home to find a gallon jar of honey on our front door step. This was his gift to us for allowing him to use the small area in our orchard.

About that time, I found an old, empty, but clean beehive in our old barn. It consisted of a bottom board, two frame boxes, all the frames, and a lid. It appeared never to have been used. At this time, Valentine was building new boxes and frames out of scrap lumber from old pallets that he cut up with an old table saw. Having no intention of keeping bees myself, I decided the old beehive would better serve Valentine, so I took it down to him the next time he was working his bees. He was delighted and started talking to the bees in Russian, telling them about the new home I had given them.

in many beekeeping books to represent typical beekeeping activity before modern times. It shows stick figures climbing up ladders and removing wild nests. Honey hunting or nest removal continues in Asia with the giant honey bee, *A. dorsata*, which has never been brought under beekeeper management.

Over the years, a developing knowledge of bee biology influenced the craft in many ways. Active management of the honey bee colony, however, awaited key pieces of information that emerged in the 18th and 19th centuries. In more recent times, innovations have increased as advancements in transportation, communication, and other technologies have come into play.

Finally, the new biological imperatives of a global community and economy have affected beekeepers in ways not dreamed of by pioneers. Several eras are, therefore, described in this volume characterizing the development of beekeeping: prehistory to 1500; 1500 to 1850, 1850 to 1984; and 1984 to the present.

Two weeks later as I was coming home, Valentine stopped me and took me to a hive away from all the others. He said, "Your hive, your family bees," and stood there looking at me.

I said, "I no know bees."

"I show," he said. It was a done deal; I was a beekeeper.

Over the next few years, with limited talking and a lot of pointing, he proceeded to show me how to work and talk to the bees. He said that the bees were part of my family and that I needed to keep them informed on what was happening in our lives. He also showed me how to take care of the hives and colonies.

Eventually I learned that Valentine's wife had put her foot down. She told him no more than 25 hives because he was spending all his free time with his bees. He gave me more and more hives until I had five. I finally told him I could not keep any more myself. I watched how he cared for and maintained the bees, and over the next five winters he did not lose one colony, so he was doing a lot right. Then he moved to Alaska and could not take his bees. He was sad but found another Russian to look after them.

I think the bees understood only Russian and were unhappy that I spoke only English. They stung me a lot, and I developed an allergic reaction. I had to quit working the hives, but my wife Lynda thought it would be fun, so she has stepped up and is doing a good job keeping Valentine's lessons alive. She is now keeping our family of bees alive, happy, and healthy.

Last year she kept three hives over winter, started a new hive from a package, and split one hive so we are back up to five families. She has joined our local beekeepers' association and is teaching some friends. I help out from time to time with the heavy lifting and with advice.

With very little spoken language, Valentine took this neophyte and showed him how to keep his bees in the Eastern European way, which has a tradition of beekeeping that goes back thousands of years. From a little thing comes a lot of fun that can spread and change lives.

Ed Carthell, Washington

This prehistoric cave painting from Cueva de las Arañas, Spain, shows how hunters have foraged for honey from time immemorial.

Prehistory to 1500: Nest Robbing and Honey Hunting

For centuries, bees were kept in hollow logs or in straw baskets. Before 1500, the height of the Renaissance, beekeeping consisted of little more than robbing honey from established nests. The Philistines dabbled in beekeeping, as did the ancient Sumerians, Arabians, Greeks, Romans, and Egyptians. Beekeeping techniques included hiving the bees and then collecting the honey through destruction of the nest. Bees were housed in a huge variety of nests, from hollow tree trunks (**gums**) to earthen pottery to woven straw baskets (**skeps**). The insects were encouraged to reproduce by **swarming**, because this was the only way to populate new nests provided by the beekeeper.

The honey bee is not native to the New World, so aboriginal Americans have no history with this insect. Certain tropical or stingless bees were kept by the Mayan and other New World civilizations, however, some using techniques similar to those employed by European beekeepers.

U.S. BEEKEEPING TIMELINE

Here is a summary of beekeeping developments, many of which occurred within a few decades in the 19th century.

1851: L. L. Langstroth recognizes the significance of bee space and builds a movable-frame hive.

1853: Moses Quinby, who also is credited with inventing the smoker, publishes *Mysteries of Bee-Keeping Explained*.

1857: Johannes Mehring produces the first **foundation**, the base on which honey bees build their comb.

1857: Comb honey production begins with W. C. Harbison of California, ushering in the "golden age of beekeeping."

1861: Samuel Wagner publishes the first issue of *American Bee Journal*. *Gleanings in Bee Culture* has also been published by A. I. Root since the 1860s; it became simply *Bee Culture* in the 1990s. (A third publication, *The Speedy Bee*, no longer in print, appeared in the 1970s.)

1865: Major F. de Hruschka invents the honey extractor in Italy.

1878: Migratory beekeeping begins along the Mississippi River.

1879: Package bees are first used.

1889: George M. Doolittle publishes *Scientific Queen Rearing*, developing the concept of commercial queen rearing based on moving larvae from comb to special queen cups — the "grafting" technique.

1926: Lloyd Watson first uses instrumental insemination of the queen bee, but it is not perfected until the 1950s.

1970s: New bee foods, such as high-fructose corn syrup and the Beltsville Bee Diet, are introduced.

1984: Tracheal mites are first detected in the United States.

1987: Varroa mites are first detected in the United States.

1990: Africanized honey bees are first detected in the United States.

1998: Small hive beetle is confirmed in the United States.

2007: *Nosema ceranae*, a variant of *Nosema apis*, becomes prominent worldwide.

2010: Large-scale "industrialized" beekeeping begins, driven primarily by commercial pollination activity.

2013: A renaissance in beekeeping arises, stimulated by concerns over the general state of honey bee health.

1500 to 1850: Evolution of Honey Bee Management

During the Renaissance and the Enlightenment, a revolution occurred in the understanding of honey bee biology. The queen was discovered to be female in 1586; drones were found to be males in 1609, and pollen to be the male part of plants in 1750; drones were shown to mate with the queen in 1792 and, by 1845, were recognized as **parthenogenic** (developed from an unfertilized egg). Although advances in biology were the rule, active human management of honey bee nests would have to wait until the development of the movable-frame hive. Meanwhile, beekeepers struggled with various techniques that were not really suitable for either themselves or the bees.

Honey bee hives came to the New World from Europe and may have arrived, first in Cuba and then in Florida, on Spanish sailing ships in the 1500s in the wake of Columbus' voyages. English settlers brought beehives to the colony of Virginia in 1622 and, soon after, to Massachusetts. Although the honey bee now seems an integral part of the American landscape, it can also be regarded as one of the first "invasive species." This efficient, social insect took the North American continent by storm. Although scientists do not really know how it affected already-established native bees, the aboriginal Indians certainly noticed a dramatic shift in their environment as Europeans also brought over a bevy of plants that honey bees pollinated.

Moving westward in swarms during the great European expansion, the bees were called by some native peoples "the white man's fly." The vast western plains may have hampered natural movement to some extent, but beekeepers helped the insects along, crossing the Midwest and the Rocky Mountains with intact colonies in prairie schooners and around Cape Horn to California by ship.

This era set the stage for the eventual domestication of the honey bee, but it was also a time when many pioneers simply took advantage of the bee as a wild or feral animal. An activity known as **beelining** became an art: honey bees were captured, given a little honey as a reward, and then released to fly directly back to their nest. The bee hunter used intersecting beelines to find the nest location, harvesting the honey and often killing the bees in the process (see box opposite).

Although many methods were devised to try to take the honey crop without destroying bees, none was truly successful. The beekeeper's constant goal was more control, but little was achieved. The biggest problem continued to be that the honey bee insisted on attaching her comb to the side of a colony. The inability to move the comb meant that beekeepers could not adequately manipulate the insects' nest or colony. They were, therefore, by necessity, what we in the craft still call "bee-havers" (see page 4).

1850 to 1984: A Golden Age

The modern beekeeping era began in 1851 when L. L. Langstroth, an Ohio Congregational minister (inspired by the writings of Dr. Jan DzierĐon, a Polish apiarist and Roman Catholic priest), discovered the significance of the "bee space," which led to the invention of his movable-frame hive. That hive became known by his name, Langstroth. Now known as the "standard" hive, it is presently the basis for most beekeeping activity in the United States.

BEELINING

Beelining is a method of finding feral nests that takes advantage of the honey bee's natural behavior. First, several bait stations are set up to attract individual foraging bees. After feeding, these insects fly in a straight line (a beeline) back to their nest, recruiting nest mates to forage at the same place. In the drawing below, we can easily find the "secret hive" through triangulation, by drawing intersecting lines from the bait stations.

Beelining is still employed in some parts of the world. Most notable is the current effort in Australia to find nests of wild Africanized (*Apis mellifera scutellata*) and Asian (*A. cerana*) bees that may have jumped ship at ports. Contemporary surveillance officers Down Under have taken a page out of the beeliner's manual to ensure that these exotics do not become established in one of the world's premier honey-producing nations.

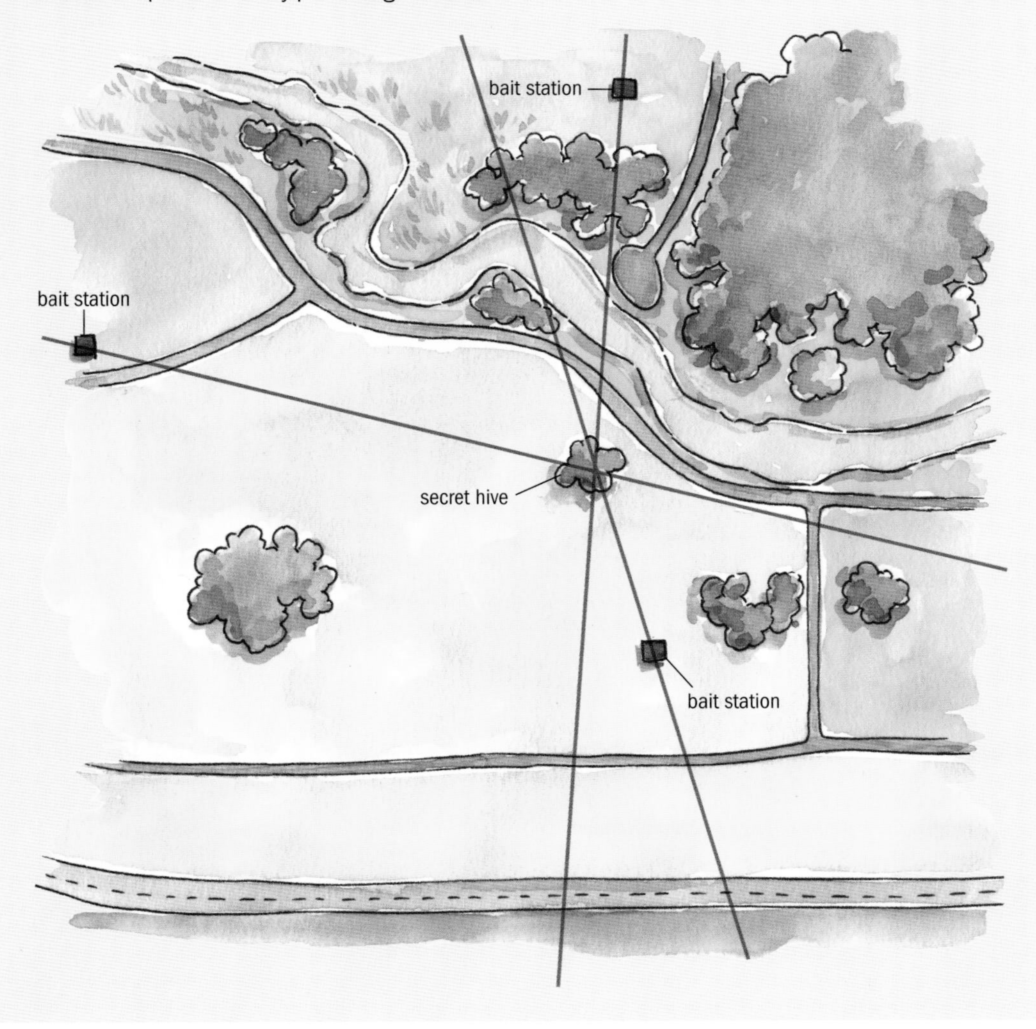

Inside a modern beehive, the frames hang in such a way that a 5⁄16-inch space is conserved between the wall and the frame edges: just enough room for two bees to pass back to back. This is known as the **bee space**, and it is significant because honey bees will neither build comb nor glue frames to the side of a hive as long as this specific gap exists. Conserving this space makes it easy to remove and replace frames. After the discovery of the bee space, beekeepers devised many techniques to manipulate beehives more efficiently.

The Industrial Revolution gradually turned beekeeping from a pastoral activity into a modern agribusiness. Larger honey crops became common, and more and more colonies were rented on a commercial basis for the fruit and nut orchards the cosmopolitan honey bee is

BEE SPACE
Honey bees will not build comb across a gap of 5⁄16 inch (9.5 mm), known as the bee space. Discovery of this led to development of the modern beehive.

adept at pollinating. Federal research facilities developed and universities and colleges offered coursework in apiculture. Advances in research on honey bee biology and the science of beekeeping continue in many of these institutions.

1984 to the Present: Modern Challenges

The year 1984 was pivotal for beekeeping in North America. Whereas the craft changed slowly up until that point, the detection of the honey bee **tracheal mite** (*Acarapis woodi*) ushered in an era of exotic organisms that would transform beekeeping in a number of ways. Quickly on the heels of tracheal mite introduction, the **varroa mite**, originally named *Varroa jacobsoni*, renamed *Varroa destructor*, was detected in 1987. Those invasions were then followed by the arrival of the gut parasite *Nosema ceranae*, as well as by virulent new strains of varroa-vectored viruses. All together, these pathogens have contributed to the condition known as colony collapse disorder (CCD).

Eradication initiatives for both these mites would prove to be ineffective. It is no exaggeration to say that a state of near panic affected honey bee researchers and beekeepers as these pests spread to almost every beeyard in the continent. Beekeepers with an anti-pesticide bias, adopted when the use of these materials by crop farmers had killed many of their colonies, quickly realized that they must now employ them to survive. This was especially the case with varroa mite infestation; if not treated it would quickly kill colonies because the insects had no innate resistance or tolerance.

Varroa—the common term used to describe the mites, the plight, and the disease affecting the bees—has so profoundly transformed beekeeping that it is safe to say at the present time the beekeeper must rethink and retool many management practices recommended in the past. The watchword now is that varroa must be controlled at all costs or the health of the honey bee colony will deteriorate.

Varroa mite management continues to be a huge distraction for researchers and beekeepers alike. Varroa is so prevalent now that, in this book, the reader will find something new and unique to beekeeping publications. In the past, only three separate individuals—the queen, the worker, and the drone—have been described as making up a honey bee colony; this volume adds a fourth, the varroa mite (see page 41).

Honey Bee Viruses Become Problematic

Varroa created another era, that of honey bee viruses. Although always present, many viruses did not have an effect because honey bees had developed adequate defenses. The varroa mite's feeding style, however — puncturing the protective honey bee skin or **cuticle**, as well as destroying the bees' critical **fat bodies** —turned heretofore benign organisms into potent killers. A few previously unimportant viruses, including deformed wing virus and Israeli acute paralysis virus, must now be reckoned with. These viruses' emergence has thus shifted our thinking about varroa population thresholds and what they mean to a colony. The mite should no longer be considered just a parasite but can be associated with a number of other maladies (see page 176).

Africanized Honey Bees Enter the United States

The Africanized honey bee was the hybrid result of accidental mating between two feral subspecies: highly defensive African honey bees brought to Brazil and previously established European honey bees. In 1990, this bee was discovered along the Texas border and has increasingly become entrenched as a population in the southern United States. More recently, it has become established in Florida, probably brought by ships from tropical America. This new hybrid has characteristics often good for honey bees (disease tolerance, swarming to disinfect the nest, superior brood production) but bad for beekeepers (over-defensiveness, excessive swarming, producing more brood than honey). For beekeepers and the general public, the balance must come down on the negative at the moment. Beekeepers face huge challenges where this insect resides because of its sensationalized reputation.

THE GLOBETROTTING HONEY BEE

As recognition of apiculture as a legitimate vocation continued, the honey bee spread worldwide with beekeepers' help. This continues even today. Significant events and dates include:

1860s: Italian honey bees first introduced into the United States.

1870s: Frank Benton imports Cyprian and Tunisian stock into the United States.

1957: African honey bees are brought to Brazil. The resultant feral bees (soon dubbed Africanized honey bees) saturate northern South America and later Central America, finally reaching Texas in 1990.

1970s and 1980s: Honey becomes a world commodity.

1990s to the present day: Commercial pollination becomes an enterprise that allows commercial beekeepers to stabilize their income and most recently has caused a boom in almond production in California.

Two More Nonnatives

Two other troublesome nonnative species have now been detected: the **small hive beetle** (*Aethina tumida*) and a variant of the traditional bee malady *Nosema apis*. The latter, *Nosema ceranae*, may have been introduced back in the 1980s but was not considered a problem until recently. There will likely be new additions to this list in the future.

Beekeeping and Modern Agriculture

The shift to larger-scale commercial pollination transformed honey bee management. This more "industrialized" model has been primarily driven by almond pollination in California, but is also recognized as important for other cash crops. It often requires a completely different beekeeping strategy than was prevalent in the past, when honey production was emphasized.

Renaissance in Beekeeping

Concern over threats such as colony collapse disorder or CCD (see page 180) has made many people conclude that the world is faced with a "honey bee apocalypse." Although the fear is exaggerated, it has increased public interest in the honey bee's well-being and survival.

One consequence is that the honey bee genome (genetic DNA code) was one of the first to be sequenced (identified and described) in agricultural organisms. The data are providing novel information useful in honey bee breeding, controlling viral infections, managing honey bees in time of significant climate change, and managing both extant and exotic honey bee pest populations. This information helps experienced and beginning beekeepers alike — those who managed honey bees "before mites" (B.M.) and those who have taken up the craft "after mites" (A.M.) — to become better beekeepers than their predecessors.

In conclusion, the intense, often sensationalized, media coverage of CCD has produced a renaissance in culturing honey bees, including a new appreciation of the insects' economic significance. Some of those reading this book may have become beekeepers as a consequence.

Onward to the Future

There is little question that the novice beekeeper faces a much more complicated task now than beginners did at any other time in the history of apiculture. Fortunately, new technologies and knowledge are being put to good use through efforts of creative beekeepers and researchers so that the activity can prosper in the future.

THE OBSERVATION BEEHIVE

The observation hive is one of the premier research and educational tools for beekeepers. It can also be used as an adjunct to a wide variety of public relations and selling programs.

Although its allure is universal, the observation beehive may not always be the best choice of exhibit. This is because a great deal of time and energy is needed to set up a hive and keep it going. Most people have few problems installing an observation hive for the first time. The next headache is maintaining the unit. This is especially true if the hive is to be used as a permanent display for the general public.

Unfortunately, there is very little that is permanent about an observation beehive without considerable work by the beekeeper. Even the largest units of four frames still only represent a portion of a full-sized colony. Because they are so small, observation hives do not usually survive major fluctuations in either population size or food availability. Anyone who has attempted to keep one of these marginal colonies for any length of time can draw up a long laundry list of potential problems. These can include: swarming, queenlessness, starvation, and invasion by diseases, pests, and parasites.

This is not to say that the observation hive doesn't have a place, only that a commitment to manage it must be made over a long time. There is nothing worse for a public display than a neglected observation beehive.

3

A BEE'S LIFE

Honey bees are wild creatures. They have never been domesticated. They have been kept, studied, researched, and bred for many years, and, in a sense, the species has been improved. But they have not been tamed. Left to their own devices, they live exactly as they have lived for thousands of years. Our true, long-term success as beekeepers comes only after we come to understand their intimate lives, behavior, and motivations.

What Is a Honey Bee?

The honey bee has been called a "flying Swiss Army knife" because of its complex structure (see illustration on page 32). It is a true insect with six legs, three body parts (head, thorax, abdomen), a hard outer skeleton (**exoskeleton**), two pairs of wings, an open circulatory system, a tracheal breathing system, and a ventral nervous system.

The bee's head contains two antennae and two compound and three simple eyes (**ocelli**). Compound eyes detect polarized sunlight, pattern, and color, while the simple eyes reveal illumination intensity (brightness). Mouthparts include paired jaws (**mandibles** and **maxillae**) and a single tongue (**labium**) that make up the sucking apparatus of the honey bee.

The **thorax** (the middle section of the body) governs bee movement. It holds the large indirect flight muscles to which the wings are attached. Six legs (three pairs) also connect to the thorax.

The abdomen contains a variety of organs and systems.

Digestive System

The digestive system consists of the mouth, esophagus, **honey crop** or **foregut** (the area where nectar is stored and transported), proventricular valve and proventriculus, anterior intestine (ventriculus), and rectum.

The esophagus leads from the mouth through the thorax and empties into a balloon-like structure called the honey crop or simply the "crop." This is the cargo hold of a field worker honey bee and can be up to 85 percent of her weight when full. This nectar never goes into the rest of the digestive tract

Honey bee workers interact with one another on comb.

ANATOMY OF A HONEY BEE

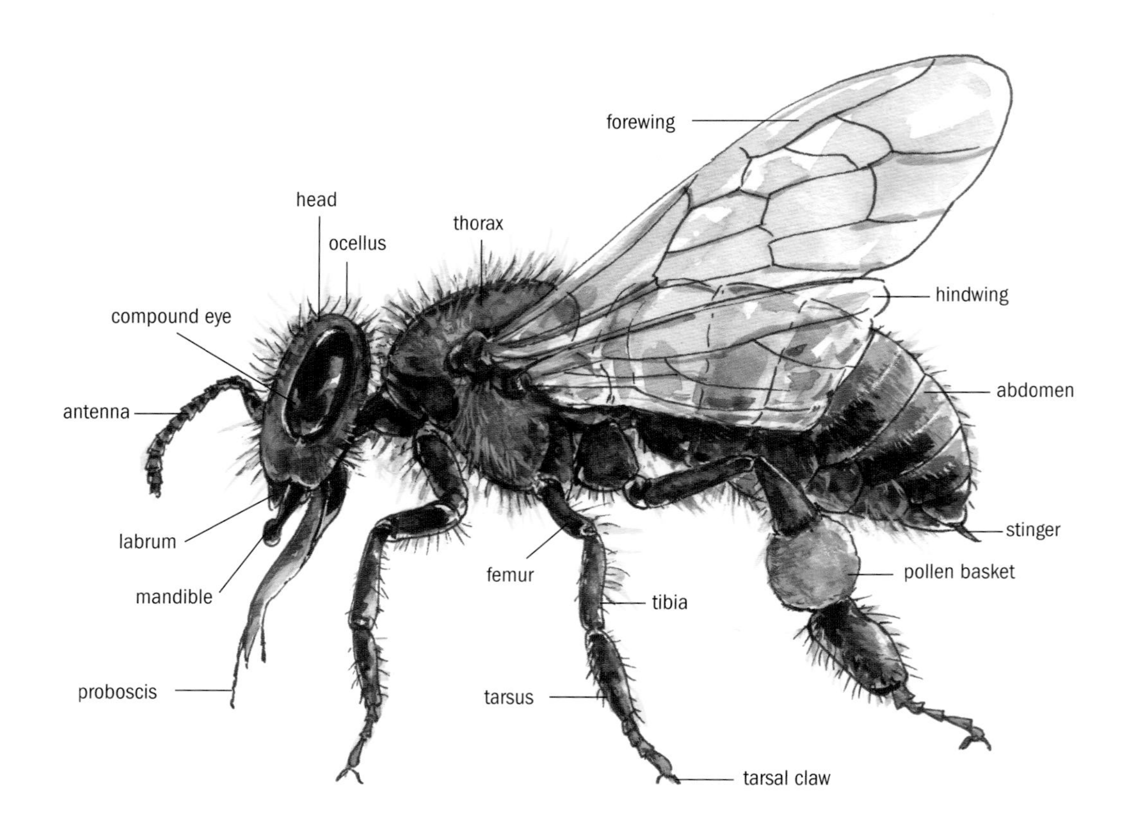

but is stopped by the ventricular valve. When a forager returns to the colony, the contents of the honey stomach are regurgitated through the mouth and given to the house bees for further processing. This means that nectar arrives in a condition very close to its natural state as it was collected in the field and so contains no possible contamination from the rest of the digestive tract.

Circulatory System

A honey bee has a dorsal heart but few blood vessels. This is known as an "open" system because the blood sloshes around inside the insect's cuticle, or skin. The blood, called **haemolymph**, is different from hemoglobin (a protein in human blood) because it does not carry oxygen, which must be acquired directly from the air. (See discussion of respiratory system, opposite.)

Respiratory System

Tubes or **tracheae** provide direct oxygen exchange between the body tissues and the outside air through exterior openings (**spiracles**). The tracheal system is perhaps best known by beekeepers because it can support a population of mites, which in high concentrations can be harmful. (See page 169 for more on tracheal mites.)

Excretory System

This system consists of small tubes, called **Malpighian tubules**, that float in the honey bee's haemolymph collecting impurities, which are discharged into the hindgut.

Nervous System

Numerous **ganglia** (structures that contain nerve cells) make up a central nervous system that begins in the head and then courses along the bottom of the body. This ventral system is characterized by paired nerve cords important in regulating muscles and organs.

An Extensive Gland Complex

The honey bee's glandular system includes **hypopharyngeal** (brood food), mandibular, wax, and scent glands. Other glands are found in many places on the body, including the feet, the sting apparatus, and the skin (cuticle). (See pages 38 and 44 for more on the functions of the glands.)

Inside the Colony

As noted elsewhere in this volume, the honey bee colony is properly called a superorganism. It is **eusocial** — a group of organisms acting together, characterized by cooperative care of the young, overlapping generations, and division of labor through a caste system. Most eusocial organisms are insects such as ants, bees, wasps, and termites, but there are some found in other groups, such as the naked mole rat, a mammal.

The Nest

In the wild or in the hive, the honey bee nest consists of parallel **combs** filled with hexagonally shaped cells. The hexagon is the structural form that provides the most strength using the least amount of material, a proven mathematical axiom. The comb is constructed solely of beeswax, produced by the bee's own

An exposed wild nest reveals parallel combs, built directly by worker bees from wax produced by abdominal glands.

body, and serves a dual purpose as both the pantry for the colony's food and the cradle for the young bees (called **brood**).

The general arrangement of the combs is for brood to be in the center, surrounded by a ring of stored pollen (**beebread**), with honey stored above and around the beebread. The outside combs have less brood and more pollen and honey, which provide insulation to maintain the necessary heat for the developing brood.

The Caste System

In the honey bee colony females do all the work, based on their caste: **queen** (fertile female) or **worker** (infertile female). The queen lays eggs and maintains colony cohesion. Workers perform different tasks depending on their age, from tending brood when they are young to foraging for nectar when older. **Drones** (males) are seasonal; their principal mission is to mate with a queen.

CENTRAL COMB showing capped or sealed honey (upper left) and a sealed brood area (lower right), separated by a middle band of stored pollen, nectar, and open brood.

Metamorphosis and Development

Especially significant for the colony is the honey bee's complete metamorphosis. Each caste undergoes a developmental process where the body radically changes in both form and function. Beginning with an egg, which hatches into a small worm (a grublike **larva**), the individual then shifts into a resting phase called the **pupa**. During the pupal process, tissues are rearranged into the final stage, the **adult**. Thus, the individual bee has specific lifestyles: eating and developing (the egg, larva, and pupa, collectively called brood); and contributing to the colony's maintenance and growth (the adult).

Development takes place in the cells of the comb. When an egg hatches, the nurse bees first offer it a nutritious jelly (**royal jelly**) in a process called mass feeding. As the larva develops, the nurses continue to provide food as needed, a process called progressive feeding. At the end of the larval period, the bees cap the cell with wax, and the larva then **pupates**, during which it reorganizes its tissues through a process called **metamorphosis** (change in form). Once this change has occurred, the fully formed adult chews away the cap of its cell and emerges to join the other adults in the colony.

The Queen

The queen is the reproductive center of the colony. She is the only female capable of laying fertile eggs, without which the colony could not grow in population. She is also the longest-lived individual, sometimes surviving several years, and the largest in size, with a long, tapering body. Because her job is to lay fertile eggs, the queen has about 160 **ovarioles**, where the eggs form (workers have four at most), and a single **spermatheca**, where sperm is stored.

Although the queen lacks some characteristics of the workers in the colony, such as wax glands, pollen baskets, and a scent gland, she is still able to sting. This helps her control rivals but is apparently not used in general defense. Humans are rarely stung by queens, but if they are, the stinging apparatus does not remain in the victim (unlike with workers) because its barbs are extremely small. Queens also have notched jaws or mandibles, which differentiates them from workers' smooth ones.

There appears to be a developmental continuum between "queendom" and "workerdom." Queens who more resemble workers are called **intercastes**. These are rare, and it is not known at what point or how a colony deals with them.

The development or metamorphosis of the queen is short when compared with that of other individuals in the colony. The egg hatches in three days, the larva pupates in just a bit more than five, and an adult emerges a week later. The total development time is 15.5 days.

DIFFERENT DEVELOPMENT RATES

Bee	Egg	Larva	Pupa	Total	Life Span
Queen	3 days	5.5 days	7 days	15.5	up to several years
Worker	3 days	6 days	12 days	21*	weeks to months
Drone	3 days	6.5 days	14.5 days	24	40–50 days

* Although the 21-day figure is commonly given, in summer weather it is more often 19 to 20 days.

Within a few days after **emergence**, the queen is ready to fly and begin the process of mating, acquiring the sperm so important to the colony's future. In the week or so after first taking wing, she must orient herself to the colony and then make several flights during which she is pursued and mated with by as many as 5 to 40 drones (averaging 15). Once she begins to lay fertile eggs, she will leave the colony again only with a swarm, when the bees move to establish another nest.

Egg production by the queen ebbs and flows with colony development and is based on many factors, including prevailing weather conditions, nutritional resources, and the makeup of the population within the hive. In the active season, some queens may lay up to 2,000 eggs a day if a large population is deemed needed, but when the conditions change, egg-laying does as well.

There is usually only one queen in a colony, but under some circumstances, two or more might be found. The usual explanation is that both a mother and a daughter — the daughter who is about to replace the mother — are present for a short period on the same comb. Occasionally, a number of virgin queens may leave with swarms, especially among Africanized honey bees.

The Worker

The worker honey bee is a fully formed female, but the effect of the queen's pheromones (queen substance) keeps her ovaries small and undeveloped. The worker population is responsible for all the well-known honey bee activities, including honey production, pollen collection, temperature regulation, brood rearing, wax and royal jelly production, and defense. There are more workers in a colony than in any other caste, numbering in the tens of thousands.

Workers are the smallest individuals in a colony and usually have the shortest life span, perhaps lasting only a few weeks in the active season. In the inactive season the worker may survive a few months, especially in temperate regions. They are also the most expendable,

queen honey bee

worker honey bee

QUEENLY INFLUENCE

Most scientists wouldn't describe the queen as "ruling" a colony, but she certainly exerts control. The most important task is to use a suite of chemicals or pheromones called **queen substance** to prevent workers' ovaries from developing to the point where they produce eggs. Workers will lay eggs on rare occasions — such as when the queen dies suddenly without a replacement — but worker-produced eggs cannot be fertilized, and they result exclusively in drones who cannot sustain a colony. Because no workers develop to replace workers who die, the colony rapidly loses population.

Another way the queen exerts control is by determining the sex of her offspring, producing either drones from unfertilized eggs or workers from fertilized eggs, so that she is responsible for the eventual population makeup of the colony. The mechanism is well understood — she either does or does not release sperm as an egg passes down her reproductive tract — but how it is accomplished and under what circumstances is not known. Because she has mated with many drones, the queen's offspring constitute a number of subfamilies of workers, some more related to each other (**super sisters** who have the same father) than others (**half sisters** with different fathers).

wearing themselves out maintaining the colony, caring for the brood and foraging for food, or giving up their lives in its defense.

The worker begins life as a fertilized egg laid by the queen, and hatches into a larva in 3 days. The larva is fed for about a week, and then pupates in 12 more days, emerging as an adult after a total of 21 days. The three-week development cycle of the worker is important to the beekeeper attempting to monitor a colony's population, either for honey production or pollination.

It takes at least six weeks to build up a colony from a maintenance population level to one that can be counted on to be productive.

Unlike their mother, workers have specialized physical structures that include a functional barbed sting (modified **ovipositor**), modified legs for pollen collection, and a scent gland. In fact, worker bees have been described as a collection of glands. These include the scent (Nasanov) gland, the **hypopharyngeal** gland (which produces royal jelly, the substance workers feed to larvae), the wax glands, and the mandibular and salivary glands (which produce important secretions involved in food production, comb disinfection, and digestion).

During their lives, workers progress through a set series of predetermined tasks, depending on age and colony need, transitioning from house bee responsibilities inside the hive to more risky activities outside as foragers.

The House Bee

The worker bee is born in the colony's brood nest, where it is warm and dark, and remains there for much of the beginning of her life. About a third of her time is spent resting and the other time patrolling the space. At an early age, her skin (**cuticle**) is soft, her muscles are not well developed, and many of her specialized glands have yet to reach maturity.

HOUSE BEE RESPONSIBILITIES

In general, most house bees go through a set number of tasks. Some of the most significant are listed here and appear in the approximate order of when they can be accomplished developmentally:

- Cleaning brood cells
- Attending queens
- Feeding brood
- Capping cells
- Packing and processing pollen
- Secreting wax
- Regulating temperature
- Receiving nectar and processing it into honey

To any casual observer, worker bees appear to move randomly around the brood nest seeking tasks they might be ready to tackle. It is truly miraculous that such seemingly chaotic behavior results in great order and efficiency. The worker may or may not exclusively dedicate herself to certain jobs.

There are both colony and genetic factors involved in determining what workers do as they age.

Tending the Nursery

Several worker tasks require the use of specific glands. The house bees feed the larvae a nutritious jelly called **brood food**, which is produced in the hypopharyngeal gland in the bee's head (the glands in dissection resemble a bunch of grapes). The gland is fully functional for only a short time during the house bee's early development when it is usually most needed. However, if the colony determines this brood food is necessary and only older workers are available, they can partially regenerate the use of the gland.

Comb Building

Later in the life of a house bee, the wax glands become prominent, reaching a peak in development at about 18 days. They appear as four discrete pairs of what are called **wax mirrors** on the underside of the bee's belly. The glands secrete a liquid that then solidifies on the **mirrors**. The bee then removes these solid wax flakes and molds them into the comb. Again, although older worker bees do not have completely functional wax glands, they can reactivate them in times of colony need.

Honey Processors

Mid-age house bees function as the colony's nectar receivers, comb builders, and honey processors. They receive relatively raw nectar from foraging bees and place it in the comb where evaporation takes place. This is facilitated by sharing the sweet among themselves through exchanging drops of nectar and depositing them in various cells around the nest. Nectar at this stage is very much like water in appearance and consistency. The workers fan their wings in unison to bring dry air into the colony and then exhaust it, full of moisture, to the outside.

During this period the bees add **enzymes** that convert the sucrose in nectar to glucose and fructose. These chemical changes, coupled with reduction of moisture in the nectar, result in the product known as honey.

The Forager

Two to three weeks after emergence, the house bee begins to fly from the colony in preparation to become a forager. These orientation or "play" flights are wonderful to see. The young bees often emerge in a cloud on a warm day and begin to fly back and forth in front of the hive, gradually traveling farther and farther from the colony until they know the surrounding landscape sufficiently well to return.

Despite this orientation period, however, many bees easily get lost, especially if there are a number of hives close by. This "drifting" behavior is not a problem in the wild where colonies are relatively far apart, but in an apiary with a number of hives, worker bees often become confused and may go into other colonies. This produces hives of unequal size and can also spread diseases and pests throughout the apiary.

Foraging is not random for field bees: they are usually guided by so-called **scouts** to the best nectar sources nearby. Once they begin foraging on a certain crop, field bees tend to be faithful to this plant, visiting it exclusively. This makes them efficient and desirable pollinators.

Foraging is dangerous. Many creatures, from predatory wasps to spiders, visit plants in bloom. Exposure to weather elements and toxic pesticides is always a possibility. The general view is that foraging bees simply wear themselves out over time. This is often noticeable in older foragers. Their wings eventually become tattered and less functional.

A forager's wings eventually become frayed by repeated contact with flowers, foliage, and other bees in the hive.

The Winter Bee

The life of the worker bee depends on climatic conditions. In temperate regions with distinct winters, in fact, there are two kinds of workers: "summer" bees and "winter" bees. Structures called **fat bodies** in winter bees enable them to survive during cold weather when colonies produce little if any brood or food. A colony's production of winter bees, therefore, is extremely important, and maximizing this population becomes a signifiant management strategy in northern areas (see Winter Management, page 119).

Fat bodies are less developed in summer bees. Scientists don't fully understand how the ratio between summer bees and winter-adapted bees is determined in the subtropics and other less temperate regions, but it is no doubt an important consequence of the different subspecies or ecotypes that have developed in specific areas.

The Drone

The drone is the male bee, a remarkable organism in his own right. First of all, he is **haploid**, the result of **parthenogenesis** or "virgin birth." That means he has half the number of chromosomes as the female. His genetics can come only from his mother because he has no father. His reproductive system is elaborate and consists of paired testes, vas deferens, seminal vesicles, and mucous glands as well as a single penal bulb and associated structures.

The drone can be considered the equivalent of a flying sperm. Most scientists and beekeepers believe his only job is to mate with a queen, providing her with a great many copies of himself in the bargain. It is possible, however, that drones perform other functions in the hive, including temperature regulation. The process of minimizing or eliminating the drone population could have unintended consequences that are not recognized by beekeepers or researchers.

The drone dies in the mating process: the wind pressure developed during his attempt to catch and mount the queen actually causes him to explode with an audible *pop* as he ejaculates inside her. He then falls off the queen, usually leaving a portion of his phallus inside her.

Research on drones reveals that in general they are unequally dispersed across the landscape. Specific congregation areas contain collections of males from throughout the area who form smaller units (called **comets**) and pursue individual queens. Unlike workers, drones are universally accepted into all colonies at almost any time.

The drone has a longer developmental period than either of the female castes. Like the queen and worker eggs, the drone egg hatches in 3 days. The larval stage is 6.5 days followed by 14.5 days of pupation, totaling 24 days. This longer development time makes drone brood

drone honey bee

more attractive to the varroa mite; one effective biomechanical control entails trapping mites in drone brood and then removing it — and the mites along with it — from the colony.

At most, only a few thousand drones live in a colony (which can number 40,000 to 100,000 bees) during the active season, and their existence is precarious. At the first sign of stressful environmental conditions, the colony disposes of its drones by banishing them from the hive. During times of nutritional deficiency, drone brood may be consumed and its protein thus recycled. The male, therefore, does have at least one other role in the colony besides mating with the queen. He becomes a dynamic nutrition bank in the active season. Some claim he is important in maintaining colony morale. Again, many beekeepers strive to eliminate drones from colonies, but this is often not successful and may be detrimental in the long run.

The Varroa Mite

Most authorities list only three individuals in a honey bee colony: queen, worker, and drone. In the last two decades, however, a fourth has forced itself into the hive: the Asian mite known as *Varroa destructor*. The varroa mite is like the twelfth man on a football team: the fans. It is not a regular player listed on the official lineup, but it cannot be ignored and can have a significant effect on the outcome of the game.

The varroa mite is not a run-of-the-mill parasite. It is an exotic organism completely new to its host, and thus the Western honey bee has little natural defense against the mite's depredations. History has proven that more than 90 percent of colonies are at risk of dying once infested. The mite is locked into the honey bee life cycle in ways not fully understood. It is totally dependent on the honey bee and cannot live isolated from its insect host for long.

BEEKEEPER'S STORY

I ALWAYS THOUGHT it would be fun to keep bees and harvest honey. I finally took a class with the Rhode Island Bee Keeping Association six years ago and my husband came along. Of course, it is more complicated than just having bees and "robbing" the honey, but what a great hobby! We work together and learn together. It has added so much to my life. I love my bees and love telling people all about these wonderful gifts of nature.

My hobby has grown into a calling to speak at schools, church groups, and libraries. I developed a PowerPoint presentation, and people are amazed about honey bees. Everyone who comes — ages 5 to 95 — is fascinated and learns something they didn't know. (Someone usually tries to play "stump the beekeeper," which is fine with me. Sometimes I can surprise them with, "Yes, I did know Sherlock Holmes was the super sleuth who kept bees," and "No, I did not know the way to artificially inseminate the queen was to give her a martini and send her out with drones.")

If you have been thinking about it, do consider trying it. The first year is a big learning curve, but it is so worth it.

Lynn Davignon, Rhode Island

Flying bees carry mites. Mites in a colony that is weakened by starvation or predation can mount a worker bee and travel with her to a healthy hive. Drones are universally accepted into colonies and can enter a large number during their lives, thus becoming a significant distributor of mites among hives. Varroa's main dispersal route, however, occurs when a colony collapses in late season, and it is robbed out by other colonies. In this way astounding numbers of mites can be carried back to the plundering hives.

The reproductive phase of the varroa mite shows how tightly its existence is woven into the fabric of honey bee colonies. Female mites are attracted to not only certain types of cells but specific ages of larvae as well. Drone brood, which takes longer to develop, seems particularly attractive to mites. The parasite has more time to produce additional mites per brood cycle than is possible in worker brood.

Because the entire mite reproductive process must take place within a certain timeframe, observers have concluded that varroa has greatly compressed its early developmental stages, actually eliminating a typical six-legged larval stage found in many other mites. Proteins from the bee's tissues also show up unaltered in mite eggs, a phenomenon known to occur only in a few other parasitic arthropods.

Our overriding reason to call varroa the fourth occupant of the honey bee hive is to convince beekeepers that the mite is not a passing phase, unlike many other diseases and pests. It is a permanent resident in all colonies in affected regions. It cannot be eradicated from the nest. This has been the futile goal of many beekeepers using increasingly toxic treatments, which are now considered counterproductive to honey bee health.

Varroa mite attached to a honey bee pupa.

Activities and Behavior

The most important aspect of understanding the activities and behavior of bees, whether it be an individual or an entire colony, is recognition that just about every bee action is attributable to some kind of situation or stimulus. The more a beekeeper understands why a bee does what it does, the better beekeeper he or she will be.

Communication

Many activities within the colony focus on communication, the basis for sociality. Honey bees use all their senses to transmit information.

Sight

One well-known characteristic of honey bees is the behavior known as **dancing**. These regular, repetitive movements communicate many things, from the location of a food source to a call for a grooming session from a nearby sister bee.

Dr. Karl von Frisch won the Nobel Prize in 1973 for his study during the 1920s and 1930s of honey bee "dance language"; he is the only honey bee scientist ever to win such an award. According to Dr. von Frisch, the dancing of scouts actually communicates four things to honey bees about food supply: its quality, quantity, direction, and distance. This has been replicated in many experiments, and humans are even able to understand these variables from a bee's dance. Another school of thought, however, holds that the dancing is not a true language, and that other factors, such as odor, may be more responsible for bees' ability to guide one another to food sources.

DANCING HONEY BEES

According to Dr. Karl von Frisch and other researchers, a "round dance" (upper drawing) indicates that a food source is near (within 100 meters). When the food source is farther away, the forager will add a straight line as a directional element, resulting in a figure-8 (lower drawing). The dancers emphasize the straight line by moving their abdomens from side to side in the so-called "waggle dance." The angle of the straight line on the comb in relation to the sun's position provides directional information.

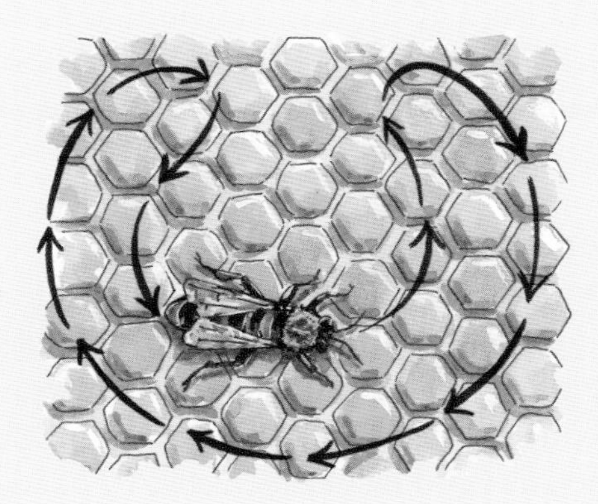

round dance

waggle dance

Sound

Sound also is important in bee communication. Research shows that the comb is involved in sound transmission and is literally a sounding board. Certain blips, often inaudible to humans, may indicate to workers the distance of a food source or the state of the colony. Queens signal (**pipe**) to each other; experienced beekeepers come to know that particular sound, just as they also recognize the characteristic "roar" a queenless colony makes.

Taste

Food sharing (**trophallaxis**) is a primary honey bee activity. It is thought to transmit nutrients and information back and forth between adults, as well as across developmental stages.

Scent

A whole class of chemicals called **pheromones**, with multiple roles, has been identified in workers, queens, and drones. Pheromones usually fall into the scent category, but also may be detected by bees in other ways. All pheromones are not necessarily volatile. A suite of these pheromones, known as **queen substance,** inhibits queen cell construction by the colony and suppresses ovarial development in worker bees. It also stimulates drones to mate with queens.

Worker pheromones also have been identified. A secretion of the Nasanov (or Nasanoff) gland recruits bees from outside the colony and orients them to the nest. Its primary components are citral (citrus odor) and geraniol (geranium odor).

An alarm pheromone component is secreted by the accessory gland of the sting apparatus, so that when a sting occurs all nearby bees are alerted. Its major component is **isopentyl acetate** (banana odor). Another alarm component is **2-heptanone**, secreted by the mandibular glands in workers, which smells to humans like blue cheese. A "footprint" substance or pheromone that allows workers to track each other while walking also has been identified.

The bees' sensitivity to odor is the reason beekeepers use smoke when manipulating colonies. It presumably masks the alarm pheromone, reducing defensive recruitment. Thus, the traditional advice when working honey bee colonies holds: Apply cool smoke from a smoker, allowing it time to pervade a colony before opening it. Manipulate the hive carefully and do not crush or otherwise agitate the insects. Avoid sudden movements and jarring

SEEING ULTRAVIOLET

Honey bees are not colorblind, but the spectrum of light that they see differs from that of humans. They can't see red, but have receptors for green, blue, and ultraviolet. These colors are often present in the nectar guides of many flowers — pathways that are often invisible to humans except under ultraviolet light. For example, the centers of black-eyed Susans, where the nectaries are located, appear very dark under ultraviolet light. That dark color, which has come to be known as "bee violet," cannot be distinguished by the human eye, but attracts honey bees.

the hive. Calm beekeepers result in calm bees. (See box, page 103.)

Learning and Recognition

A large body of literature focuses on how intelligent honey bees are and whether they are adept learners. The answer to the former question is not clear; intelligence is a subjective term that involves anthropomorphism more than science.

On the other hand, it is clear that honey bees can learn. This has been proven by a variety of experiments, many using the same principle of **behavioral conditioning** (learning new behavior based on specific previous experience, positive or negative) that Russian scientist Ivan Pavlov used with dogs. In honey bees, Pavlovian behavioral conditioning is easily observed in the honey bees' proboscis extension response (PER), and they have been trained to recognize all kinds of substances, from drugs to explosives.

HONEY BEES AS DETECTIVES

Scientists can now train honey bees to detect unique chemical molecules. A remarkable 95 percent accuracy rate, compared with 71 percent in bomb-sniffing dogs, has opened up the prospect of using bees to detect improvised explosive devices (IEDs), suicide bombers, biological weapons, and eventually diseases and cancers in people. Most exciting is the real possibility of using honey bees to detect the many plastic land mines that populate the earth and cause a great deal of human suffering.

Honey bee training uses classical Pavlovian behavioral conditioning, focusing on the proboscis extension reflex (PER). At least one device attempts to translate honey bee behavior into an electronic signal that can be detected on a screen.

Bee Emotions?

Whether honey bees experience sadness is not known, although there is an Old World custom of "telling the bees" when their beekeeper has died. And although there is no way to tell what bees themselves feel, some believe that, like dogs, bees can detect fear in humans.

It is often said that honey bees can recognize their beekeeper. Pattern recognition has been detected in honey bees; it is possible that they become attuned to a specific human face. It seems unlikely that individual bees get to know any one person, however, because the average beekeeper doesn't visit a specific hive often enough. An individual worker honey bee may live only four to five weeks in the active season, so there is not much opportunity for exposure to a single human being. More likely, the bees focus on the behavior of a human beekeeper.

One other possibility is that chemistry is involved. Many people attract the interest of bees because they routinely wear perfume, scented deodorant, or hairdressing. Some animals, such as horses and dogs, are known to elicit defensive behavior in honey bees far more than other organisms. It is thought this is the result of particular odors, behaviors, or some combination of both stimuli.

Patterns of Behavior

Honey bees show many behavioral patterns that beekeepers must take into consideration when managing colonies. An example is the colony's reaction to smoke, a major beekeeping tool. When a colony is actively smoked, bees first move away from the source and then begin to engorge themselves with honey from the comb, which makes them less likely to sting.

Other patterned behaviors include temperature regulation (**thermoregulation**); swarming, which is part of the reproductive process; and **absconding**, or migration, in warmer climates. All of these directly affect the well-being of a colony and its productivity. The best beekeeper recognizes these patterns, understands their significance, and stands ready to react should it become necessary.

Temperature Regulation

Unlike mammals, honey bees are unable to regulate their body temperature for extended periods of time. The thermoregulation of social insect colonies, therefore, is a great advance in survival strategy with the following advantages:

- A honey bee colony can be active year-round, even when temperatures are extreme
- In temperate areas, honey bees can produce more brood earlier in the season than bees with annual life cycles (bumble bees)
- The honey bee can both heat and cool its nest, keeping the brood nest temperature at a consistent 95°F (35°C)

Clustering for Warmth

During cold weather, individual worker honey bees shiver or vibrate their flight muscles to create heat. They do this while huddled together in a ball or "cluster." The older bees line the outside of the cluster to insulate it from heat loss. Younger bees are passively heated inside the mass of bees. A steep temperature gradient often forms between the cluster's center and the outside, and the ball of bees expands and contracts as the bees move farther apart or closer together in response to the surrounding air temperature. The cluster of bees forms when the temperature outside drops down to about 57°F (14°C) and disassociates at higher temperatures when heat creation is no longer necessary for the hive.

Cooling

Honey bees cool the colony by collecting water and fanning air currents through the colony. During hot weather, a large number of bees may also hang outside the hive at night, seeming to drip off the bottom board. This mass of workers is known as a **beard**, and the behavior is called **bearding**.

Although beginners often fear this is a sign that swarming is imminent, that is not the case; it is a normal response to high temperatures, especially when coupled with high humidity.

Beekeeper Management

Beekeepers can help colonies regulate temperature in a number of ways. In colder climates, you can paint hives black to absorb heat; wrap them in various materials, such as straw and insulation; and place them in sheltered areas. A beekeeper's help can only go so far, however, as the insects really do not heat the total interior of a hive, but only the discrete cluster of brood.

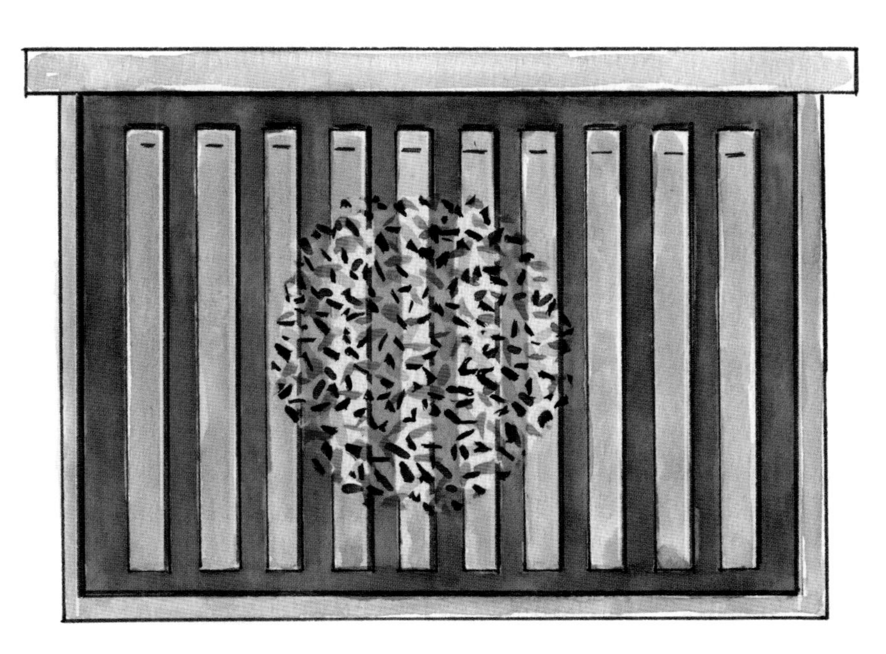

A cluster of honey bees within the hive producing and conserving heat in cold weather. (See page 46.)

A beard of bees clusters on the front of a hive on a September day.

In hot regions, shade and white paint are recommended. A spray of water can also help cool a hive or an overheated cluster of bees, especially during transit.

Swarming

Swarming is a process by which one colony of bees becomes two. Why a particular colony swarms at a particular time is not clearly understood, but in general, the colony is reacting to factors such as nectar flow, numerical strength of the colony, age of the queen, or perceived congestion in the hive. Swarming can be seen as a compelling force that takes over the colony some days before its actual division and the subsequent departure of about half the population as a "swarm."

When nectar is abundant and as the population builds, subtle changes begin to occur in the hive, triggering the swarming impulse. Crowding and congestion result in an excess of brood food because there are too many house bee nurses for the number of larvae present. Critically, there may not be enough queen substance produced to counter ovarial development, resulting in workers laying eggs. These unfertilized eggs develop into drones that cannot support the colony.

A MATTER OF PHYSICS

It is often asked why a colony dies during winter in a temperate zone (see pages 104–106) even though plenty of stores fill the brood combs and supers. This is especially frustrating to those who carefully obey the maxim to leave the bees enough stores for winter. The size of the cluster of bees that bunch together to survive cold weather is all important in wintering. If the cluster is too small, the population is not efficient in producing and conserving heat. In terms used by physics professors, the cluster's surface-to-volume ratio is inadequate.

The fall bee population is made up of winter bees who are programmed to save energy. Any reduction in brood rearing in autumn translates into fewer winter bees. A certain number of these specialists is needed to produce and conserve heat in the cluster and begin the arduous work of population buildup the following spring.

A late honey flow can be a major reason for a small cluster. In their frenzy to get the last drop of a bumper fall nectar crop, the bees may pack the brood nest with honey, and the queen's egg-laying space can become restricted. She becomes "honey-bound," with no room to produce the necessary brood that would develop into vital winter bees, ensuring a cluster with an adequate surface-to-volume ratio.

The winter cluster moves over time, covering and incorporating stored food reserves. It moves upward, as the insects store most of their food above the brood nest. The cluster size, therefore, must be flexible enough to be able to move. If it is too small, it risks being stranded like an island in a sea of honey it can't get to. Too large a cluster may mean the bees eat themselves out of house and home before winter concludes. Thus, an optimal surface-to-volume ratio is critical for any honey bee colony to survive the winter. This is not as critical where severe cold is not an issue.

As a colony prepares to swarm, nurse bees raise new queen candidates in specialized peanutlike queen cells.

Preparing for Departure

The workers' response to these signals is to reduce the queen's food supply. She is being readied for flight and is put on a diet.

Meanwhile, the colony begins to construct queen cells, encouraging the nurse bees to switch into a queen-feeding regimen. The cells are completely different from others found in the colony. Resembling peanuts in the shell, these are usually found along the bottom edges of the combs (as in the photograph above). One of the residents will ultimately be the replacement for the old queen, who will soon leave the colony.

Taking Off

When everything is ready, about half the colony departs, usually early on a sunny day. Which bees go and which stay remains a mystery. The swarm leaves with the old queen and generally moves to a temporary bivouac. This is the mass of insects the general public sees; all too often it is a source of alarm. The panic is totally unjustified as swarming bees are not the least bit interested in stinging. They have lost a home to defend and are in search of another and seem to understand this.

Collectively, the old home location appears to be forgotten by the departing occupants, and scout bees set out in search of appropriate new homes: hollow trees, cavities, and walls of buildings. The scouts report their findings by dancing; a collective decision is then made to take up lodging in a new place. If no decision is made, the bees may begin building comb on the spot of their encampment.

On the Home Front

Meanwhile, back in the original colony, a young virgin queen emerges from the queen cells left behind. She must exit the hive, mate, and return safely, presenting huge risks to herself and the colony. It is, therefore, known that workers in swarming colonies often hedge their bets by keeping some young virgin queens trapped in their cells in reserve until it is determined that replacement has been successful. The process is not perfect, and this is why beekeepers may find multiple queens in a colony, including a former-queen mother and her virginal daughter, or multiple virgins. In the end, however, it is all resolved, and one queen again is selected as the reproductive center of the colony.

There is little hard information on what actually triggers swarming, but many think congestion and an abundance of sealed brood are main stimulants. Factors that contribute to a sense of crowding are lack of ventilation, inadequate room for nest expansion, and a shortage of queen pheromone.

Beekeeper Management

Many experienced beekeepers have been frustrated because swarming can set back the parent colony, cost a honey crop, and in some cases even threaten the survival of the unit. Once the swarming impulse begins, there is little a beekeeper can do to thwart the process. (See page 95 for more on swarming.)

Absconding

In temperate regions, honey bees are not migratory and depart from their hives mostly as a reproductive strategy. Bees in those areas are more concerned with storing honey for winter survival, and they swarm only in times of great excess.

The behavior shown here is **washboarding**. The bees seem to be scrubbing the hive front by repeatedly moving up and down. It is common in hot weather, but not necessarily a sign of imminent swarming.

Honey bees in tropical areas, on the other hand, often need to migrate in search of water or to avoid infestations of diseases and pests. They often **abscond** (leave the nest altogether) in response to severe environmental conditions. When a colony absconds, all the bees leave, whereas 50–70 percent leave during a swarm.

Although reproductive swarming and absconding are very different phenomena, to the beekeeper and general public they appear to be the same, since both result in a group of honey bees departing in search of another nesting site. Absconding rarely occurs in colonies in temperate regions where European bees are usually kept, but it may occur if colonies are under threat from disease, pests, or depletion of forage (pollen, nectar, or water). In more tropical regions of the Americas and Africa, honey bees abscond much more frequently. This is one reason the Africanized honey bee is described as one variety seen swarming so often.

Robbing

Robbing is another of those anthropomorphic words used by beekeepers to describe a behavior often seen in apiaries, where bees from one colony enter another and attempt to "steal" stored honey. From a different perspective, this behavior can be seen as a specialized kind of foraging, one not usually found in nature, where colonies are not situated as closely together as in apiaries designed by humans. (See the box on page 56.)

In a robbing situation, the general excitement of bees in an apiary increases; many may be seen hovering around hives as if they are "checking" them out. And they are! Any crack in a colony or breach of defenses provides an opening for robbing to begin. Guards at entrances and other openings are also much more agitated. In serious situations, stinging (including people and animals near the apiary) can become epidemic.

Honey bees are attracted to neighboring colonies during dearth periods. If they can gain entrance to a colony, they will take honey from that hive and attempt to carry it to their own. Most colonies have a cadre of guard bees to prevent this. In some cases, however, the defenses are breached. When this happens, other bees are recruited from nearby colonies, and chaos ensues as stinging becomes rampant.

Beekeeper Management

During nectar flows and when conditions are optimal, the beekeeper has the luxury of inspecting colonies when opening a hive provokes only a minimal amount of robbing and defensive behavior. During times of nectar dearth, however, robbing among colonies coupled with stinging can become a real problem. The beekeeper must guard against this at all times by not leaving either honey or syrup exposed to foragers. If a robbing situation gets out of control, there is little the beekeeper can do but retire from the scene, plug the entrances of colonies so that only one bee can enter and exit the hive at a time, and let the behavior run its course.

4

CHOOSING HIVE LOCATION

For beekeepers, as for real estate brokers, much of the time success boils down to "location, location, location." It is also well to keep in mind another beekeeper's maxim, paraphrased from the political realm: "All beekeeping is local."

Determining where to locate a hive is often obvious because of limitations, but in other situations, many options may be possible. As in all things, it is best to look at potential sites from the bees' point of view.

Studies done in temperate locations, where nests are usually found in hollow trees, reveal certain natural preferences. The insects choose an elevated location, about 10 feet off the ground, with a southerly exposure and not too much sun or shade. In some tropical conditions, however, trees are not available and bees may have to settle for suboptimal conditions. Aside from the bees' preferences, the beekeeper must consider accessibility and the potential apprehensions of neighbors, all the while applying a measure of common sense.

Four Tips on Hive Location

Out of sight is out of mind. The closer your hives are to your residence the better. Beginners need colonies nearby so they can visit them often. Casual visits are important even if you don't open the hives. You can learn much simply by observing the entrance.

Don't invite vandalism. Isolated beehives can attract the attention of vandals, vagrants, bored kids, and thieves. Colonies have been stolen in their entirety, pelted with rocks, tipped over, targeted with gunshot, and tossed into rivers. Keep them where you can see them easily, but camouflage your hives behind shrubbery and/or locate them behind fences and buildings.

Provide some room to work. It takes space to manipulate colonies, so do not locate them too close to one another. Always ensure there is enough space between colonies so they can be worked without disturbing those nearby.

Stay level. Colonies must be on level ground. Honey is heavy and moving it is often no fun; the beekeeper should use aids like wheelbarrows, hand trucks on inclined planes, and mechanical lifts to transport filled supers.

This beehive, on an old farm in rural Massachusetts, produces far better than its in-town counterparts. Its optimal location features a forest windbreak on two sides, lots of sun, and plenty of forage. Note the bear fence around the site.

ROOFTOP HIVES

One of the best locations for honey bees is a roof. Many apiaries thrive on city rooftops from Paris to Chicago. The benefits to a rooftop location include the fact that your hives avoid interference by humans and other animals. And they are easy to monitor. Urban bees forage in parks, gardens, street trees, and balcony flower boxes: any nectar source within approximately 1 mile of the colony is potentially available. City bees in some cases can be more productive than country bees.

The challenges of keeping bees on the roof include the difficulty of lifting honey-filled supers by hand, since you can't use trucks or other big mechanical devices. In addition, a roof can be sweltering in summer and/or extremely cold in winter, when you'd rather not be out on the roof yourself! Fortunately, bees are good at maintaining their own temperature. Be sure the lids are firmly held in place in case of high winds.

Beehives on the roof of Notre-Dame de Paris

The Colony and Your Community

By establishing just one beehive, you are introducing tens of thousands of insects into your community. This is bound to have an environmental impact, even if subtle and benign, and it's important to think about the effect on the other residents — human and animal — of your neighborhood.

Consider the Neighbors

Urban beekeeping brings into focus a number of challenges. Neighbors are a constant concern for beekeepers. The beginner must avoid becoming the victim of those who see any insect as a nuisance, especially one that stings. This goes in spades for anyone keeping bees where Africanized honey bees have become established. The following are some suggested procedures to avoid potential problems caused by beekeeping activity.

Look up local regulations dealing with keeping bees or any animals in your neighborhood or area. If there are intolerable restrictions, consider writing up a specific ordinance for your situation. Many local and state beekeepers' associations will help in this regard, and some good model ordinances are also available. (See page 200 for a sample.)

Place colonies away from lot lines and occupied buildings. If near buildings, locate hives away from entrances and lines of foot traffic. Hive entrances should face away from heavily traveled areas.

Erect a 6-foot barricade between the hive and the lot line to force the bees' flight path upward. Use anything bees will not pass through, such as dense shrubs and fencing. Anytime bees are flying close to the ground and across the property line of a neighbor, there are potential problems.

Provide a water source. If there is no natural water source nearby, and especially if swimming pools are in the vicinity, place a tub of water in the apiary. Add wood floats to prevent the bees from drowning. Change the water periodically to avoid stagnation and mosquito breeding. Honey bees often drink from chlorinated swimming pools, which won't harm them but can disturb humans in the vicinity. Honey bees can be effectively trained to use urban watering sources when you add a tiny amount of an attractant such as lemongrass essential oil or, even better, a little bit of salt, tree bark, or aquatic plant life to the water source.

Minimize robbing by other honey bees. Robbing bees are usually quite defensive and will be more likely to sting passersby. Manipulate the hives only during nectar flows, if possible. Avoid exposing extraneous honey or sugar water or robbing is likely to begin. (See box on next page.) Use entrance reducers to minimize the likelihood of stronger colonies robbing weaker ones.

Prevent swarming. Although swarming bees are the most gentle, a large, hanging ball of bees often alerts neighbors to beekeeping activities and may cause undue alarm.

Keep no more than three or four beehives on a lot less than 0.5 acre. If you desire more colonies, find a nearby farmer who will host bee colonies in exchange for some honey.

Open your hives when neighbors are not in their yards.

Requeen overly defensive colonies.

Give a pound or two of honey each year to your neighbors.

Reducing Defensiveness

There are a number of beekeeping techniques that may reduce the defensiveness of a colony. Some involve location; others require a deeper understanding of honey bee needs. Here are some principal methods:

- Avoid placing too many colonies in one location.
- Locate hives in direct sun.
- Always use smoke when inspecting bees.
- Keep the area around the hive free of exposed honey to minimize robbing.
- Avoid moving hives.
- Do not open hives during bad weather.

ROBBING

The beekeeper must continually guard against robbing for a number of reasons. Most important is that this is a major way bacterial diseases are spread. Robbing can also lead to severe stinging incidents. The excitement level shown by honey bees engaged in a large robbing episode is something not soon forgotten. Whole apiaries have been reportedly destroyed in some instances.

Some robbing probably goes on all the time among colonies. This endemic robbing level only becomes epidemic when conditions are right.

Beginners often unwittingly allow robbing to build up when they keep colonies open for long periods while making inspections or taking off honey. Weak colonies are particularly vulnerable. The best way to protect such colonies is to reduce entrance openings with wooden blocks, grass, or straw. In queen-rearing operations, specialized "robbing screens" are used to protect small colonies or nuclei from invasion.

Once robbing has become epidemic, it is almost impossible to stop. About the only thing a beekeeper can do is put colonies back together and reduce their entrances, as well as those of nearby colonies, letting the behavior run its course.

When entering an apiary, always take stock of the robbing potential by maintaining a constant awareness of the bees' excitement level. If robbing gets out of hand, both bees and beekeeper suffer. Controlling this destructive behavior is one of the quintessential acts of good beekeeping.

Consider Domestic Animals

In rural areas, ensure that livestock are not located nearby (some tend to use beehives as scratching posts). Horses are a particular problem. Their odor seems to offend bees, and when stinging starts, horses often react violently and can injure themselves.

In addition, horses are not as tolerant to venom as humans and can easily be killed. Realize that horses are some of the most potentially valuable livestock, and a stinging incident could incur a major legal risk.

Wild animals that might investigate beehives include bears and skunks.

Providing Water

Many areas where bees are located may experience dry times during the course of the year. When intermittent creeks cease to flow and tree leaves show signs of moisture stress, bees become more noticeable to the general public. This can add up to telephone calls about honey bees collecting water from leaking faucets, bird baths, pet dishes, and swimming pools.

The beekeeper must provide a water source for bees if there is any likelihood the insects will forage in nearby urban areas during dry spells. Prevention is the only cure for this problem. Don't let the bees become trained to a watering place like a swimming pool. Once a water foraging pattern has been set, it is almost impossible to do anything to change it.

Locating bees near accessible water is the best way to provide a continuous supply. It is also important to make sure that any potential water supply is not contaminated. If no source is located nearby, providing water in the apiary is possible, but often requires a good deal of planning and thought.

Fill 55-gallon barrels or other containers with water and layer wood floats on top to keep the bees from drowning. A problem with this kind of device is potential stagnation. Probably the best device is one that trickles water down a wooden board or slowly drips onto an absorbent material, keeping the surface damp.

Weather Issues

Take into account the climatic elements in your region as you choose your site. In desert conditions, shade and wind breaks are necessary; not so in forests, where too much shade can be a problem and might result in high humidity in hives. There is evidence that bees in direct sun are not as defensive as those in shade.

Windy sites can be challenging for bees. They do not fly well in gusts of more than 15 miles per hour and can become disoriented. Strong gusts may damage or topple tall hives.

Water availability is extremely important, as already described.

BEEKEEPER'S STORY

I NEVER BUY PACKAGES, especially from out of this state or region. Rarely do I buy a nuc; mostly I raise and split my own hives. I raise several dozen queens every year, and I run two Carniolan, one Italian, and two mixed-mutt-survivor lines. One line is from a hive taken from a house during a removal. (I collect swarms and remove bees from houses, structures, and trees.) That queen proved to be very successful during her first year, so I started breeding her line in the second year. Now, in the third year, she is still doing well. I evaluate a new line for at least one year, preferably two, then start breeding.

It's difficult to get beekeepers to understand that bringing in bees from other regions (or even worse, other countries) is damaging our industry. Evaluating, selecting, and raising local bees and queens is the most sustainable way to keep our bees healthy. Bringing new species or strains from other regions has been a mistake that human beings have repeated time and time and time again. When will we learn from our mistakes?!

Remember that bees have lived in cavities for eons without our help. They do not need us. Our job is not to control or "fix" the bees. Our primary responsibility in this partnership is to assist only as much as they need.

The typical configuration here is two deeps or one deep plus a medium. A few people use only mediums. This area is mostly forest and pasture land so, unlike Indiana, where I used to keep bees, the major nectar source here is white Dutch clover. The *Rubus* species (blackberry and wild raspberry) also count heavily. The climate here is very hot and humid during summer, but it dries out in July and August. The winters are mild. The only real nectar flows March through mid-July. In Indiana, I had nectar flow from March through mid-October.

It has been a real learning curve down here. It is not a really good beekeeping area.

Jeffrey Maddox, Missouri

Forage Availability

Bees cannot make honey or survive without proper forage. Fortunately, the insects are cosmopolitan and can take advantage of a wide range of plants. Many urban areas support honey bees, but the beekeeper often has no idea what plants the insects might be foraging on. The best source of information on this is the local beekeepers' association.

Most locations can support a few colonies, but as the number of hives increases, pressures on the local plant resources increase. It is the local area within a one- to two-mile radius of a colony that counts.

Locations with adequate forage are difficult to find and keep, and beekeepers tend to guard them jealously. It takes several years to determine whether a site will be good on an average basis. One year's bumper crop can be followed by one or two years of flops and vice versa.

There are a huge number of plants that honey bees might visit, but only a few are considered major nectar sources in any specific region. Many books have been published on major and minor honey plants and their ranges (see Resources).

LOCATING HIVES NEAR PLANTS

If pollination is a goal, realize that nearby wildflowers may be more attractive to honey bees than those you want pollinated. Cucurbits like squash and cucumbers, for example, might need honey bees for pollination, but this goal is easily thwarted if more attractive plants are in bloom in nearby woods or fields. (See chapter 9 for more on pollination.)

Forage availability changes with the times. There are many examples of beekeeping operations that have had to move because of large-scale forage loss. For example, after World War II, beekeepers had to change locations from Alabama to Florida when their traditional nectar source, the tulip poplar (*Liriodendron tulipifera*), was cut down to aid the war effort. On the other hand, cotton honey has recently begun to flourish across the South because of the success of the boll weevil eradication program. Before this occurred, locating honey bees in cotton country was a death sentence because of the extensive use of pesticides.

In Ohio, one can draw a line down the middle of the state from north to south. On one side, sweet clover honey yields much more than on the other because of the underlying limestone rock formations. And in the Panhandle of Florida, the Chinese tallow or popcorn tree (*Triadica sebifera*), a nonnative invasive, has recently become a major nectar source, whereas before it was practically ignored by beekeepers. The famous white tupelo tree (*Nyssa ogeche* Bartr. ex Marsh.) produces large quantities of nectar only in a localized area around the mouth of the Apalachicola River in Florida. In many regions of Europe, the black locust (*Robinia pseudoacacia*) is a major honey plant, a much more reliable source of nectar there than in its native range in the eastern United States.

Many of the plants honey bees forage on were brought over from the Old World. They transformed the landscape of the Americas, and the process continues today. Introduced plants are huge problems for some areas, but honey bees often like them just fine. Perhaps the best examples of this are in Florida. Two plants in particular, Brazilian pepper (*Schinus terebinthifolius*) and melaleuca or "punk" tree (*Melaleuca quinquenervia* [Cav.] S. T. Blake), bloom in September and October when honey bees need nectar most. They are extremely invasive, however, and are blamed for many ills, from drying up wetlands to choking out native species. Beekeepers have sometimes been at odds with those who would like to reduce the footprint of these plants on the landscape.

Global Climate Change

Another consideration is that global climate change appears to be influencing local nectar flows. A NASA climatologist in the Maryland area has recently examined **hive scale** data (recording the weight of the hive through the year), which verifies that nectar sources are blooming earlier in spring each year and, ironically, also later in fall. The Bee Informed Partnership or BIP (see page 205) is mounting a project to increase the number of beekeepers using hive scales as a way to monitor shifts in **phenology** (the blooming times of plants). (See pages 75 and 117 for more on weighing hives.) Other technologies that might help beekeepers in locating suitable bee forage in the future include satellite remote sensing and global positioning systems (GPS).

CONSERVING BEE FORAGE

Beekeepers should recognize that only a few of the many existing blooming plants are major nectar producers, although most will provide at least some nectar and/or pollen to bees at certain times of the year. The vast majority of important bee plants are feral or wild in nature. This makes beekeeping, like fishing, an extractive industry. It is based largely on using plants that are not cultivated and are subject to many influences outside the beekeeper's control.

In the Midwest, corn and soybean plantings have led to the cultivation of marginal lands, previously reserved by default for many nectariferous plants. And in the South, urban sprawl and extensive drainage of land for agricultural purposes continue to reduce land inhabited by plants beneficial to honey bees.

There is no unanimous agreement on how to solve the dilemma of decreasing nectar resources. Approaches include:

- Large-scale agricultural planting of nectar-secreting crops
- Sowing roadsides, reclaimed mining areas, and fallow land with nectariferous vegetation
- Mowing roadside vegetation instead of treating with chemicals
- Developing plants that are superior nectar producers through genetic engineering
- Introducing species that are proven nectar producers in other regions

It will take cooperation among a number of parties involved in maintaining and reestablishing the feral plant landscape to conserve even a modest amount of honey bee forage in the future.

POPULAR NECTAR SOURCES

Tulip poplar
Lirodendron tulipifera

Chinese tallow or popcorn tree *Triadica sebifera*, also referred to as *Sapium sebiferum*

Tupelo
Nyssa ogeche

Black locust
Robinia pseudoacacia

POPULAR NECTAR SOURCES (CONTINUED)

Brazilian pepper
Schinus terebinthifolius

Mesquite
Prosopis spp.

Goldenrod
Solidago canadensis

Dandelion
Taraxacum officinale

Cotton
Gossypium hirsutum

Blackberry
Rubus fruticosus

Yellow sweet clover
Melilotus officinalis

5

GETTING EQUIPPED

Now that a number of preliminaries are out of the way, it's time to consider some of the first decisions novice beekeepers must make with respect to beekeeping equipment. The considerations in this section are the same for anyone keeping bees, but may vary depending on the beekeeper's objectives and experience.

Beekeeping equipment and supplies are available around the world. In the United States, several supply houses have been open since the 1880s. New ones crop up occasionally. Some beekeepers make it a pastime to collect catalogues and compare prices. The Internet has now made many of these firms easily accessible wherever you live.

The array of items available to the beekeeper is enormous and can be overwhelming to the novice. One of the beginner's challenges is not to get carried away by all the options. A good rule is to stay with equipment that is consistent with your objective; never buy an item without understanding its value and potential use. This book will stick to the basics, encouraging the beginner to start the craft in as simple a manner as possible.

A number of other components can enhance the standard hive body, including a bottom board, a honey super, a hive stand, a queen excluder, and a feeder. In addition, specific tools all beekeepers need in their arsenal include a smoker, a hive tool, and protective clothing with a veil to cover the face.

Hive Design and Dimensions

The movable-frame beehive is the foundation of modern beekeeping. This cannot be overemphasized. Most managed beehives around the world are of the type pioneered by the Rev. L. L. Langstroth. The dimensions he used, a box accommodating 10 frames, 9⁹⁄₁₆ inches high (often called **full depth**), have become standard in the United States. The use of "standard" equipment, therefore, is the cornerstone of modern beekeeping in the United States; as a consequence, most equipment available to the novice in this country is interchangeable, manageable, and readily available. Beekeepers not using standard or Langstroth equipment risk future problems when they expand or combine their operations with someone else's.

Holding bee brush and hive tool, a beekeeper in full regalia, including veil, suit, boots, and gloves, inspects a comb.

ANATOMY OF A BEEHIVE

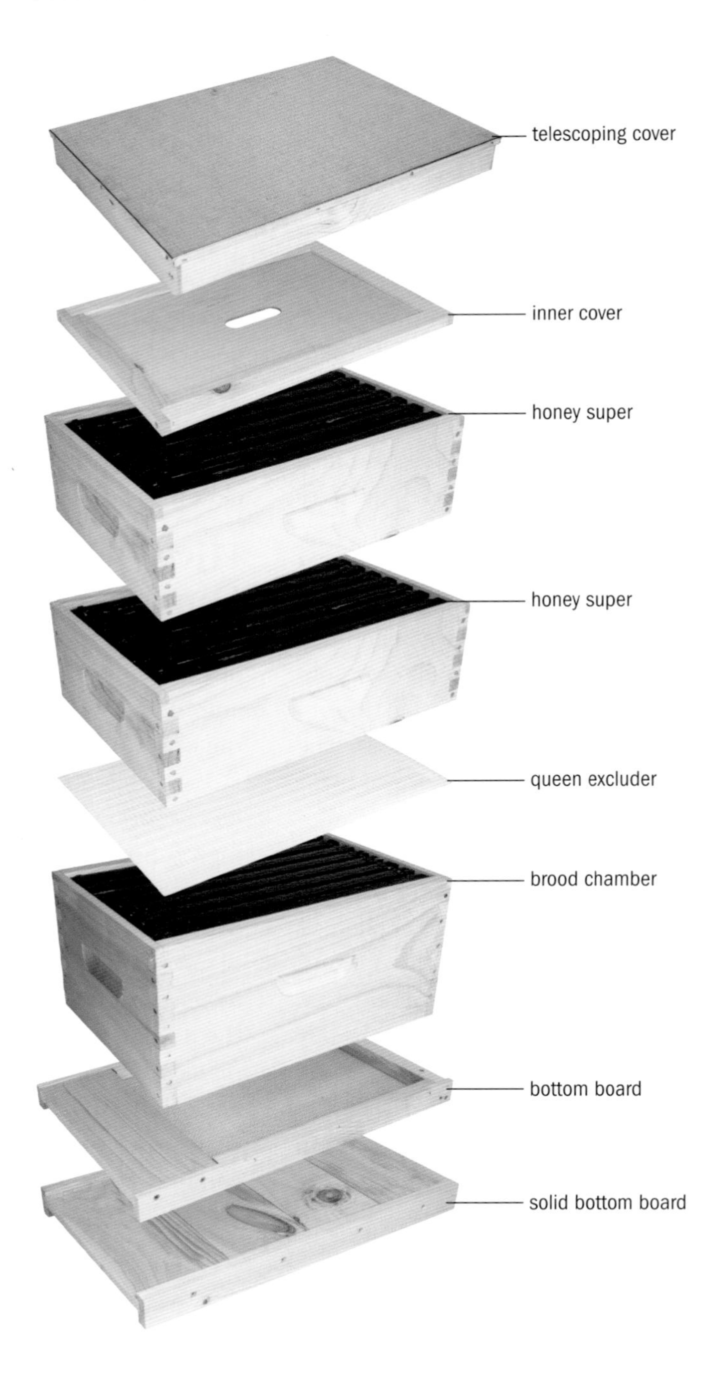

Given that advice, beekeepers should know that other systems are also available and make sense in some situations. These include hives based on eight frames, which allow for easier lifting for the operator, and the shallower (6⅝-inch) Dadant square box, for those wishing to interchange equipment extensively. Another popular system is the top bar hive, a radically different design employed by some beekeepers who prefer a small-scale approach to the craft (see pages 190–193).

Hive Materials

Traditionally, the standard box or hive body is constructed of wood, although other materials such as plastic and even concrete have been used in some areas with success. Beginners should stick to tried-and-true standard wooden equipment, and experiment with other configurations only as they gain confidence and experience. Most hive bodies are made of pine or cypress, both of which will provide years of service.

Exterior Treatment

Many authorities recommend painting the exposed surfaces to lengthen the box's life. The majority of hives are painted white, which allows hives in full sun to stay cool; however, it also makes a colony much more visible in the landscape. In cold climates, black or other colors may increase heat absorption from the sun.

The bees don't seem to care what color a hive is painted. It is possible, however, that different-colored units may help foragers find their specific colony in a crowded apiary. Cheap sources exist; some paint stores and recycling centers give away off-color or leftover paint.

Some beekeepers treat hive bodies with a preservative, but this is risky as bees can be affected by all kinds of wood treatment chemicals, which are often insecticides in their own right. An alternative is to use paraffin wax dips in various formulations.

BEEKEEPER'S STORY

I HAVE THREE LANGSTROTH HIVES, one of which is a very recent split. I would like one more hive, a vertical top bar hive of the Warré design. Four is plenty. (Three is actually plenty, but I really want that top bar hive. I do have a life outside of beekeeping.)

I bought nucs for the first two hives, one of which subsequently swarmed then made itself a new queen. We made a split from these two hives last week. I bought the queen for it in north Georgia. So far I've had Italians. The new queen is an unknown conglomerate with hopefully desirable traits. We shall see. I suspect I will see varroa mites any day now.

Having a mentor close by to guide me, especially hands-on when I'm in the hives, would be so valuable. It would ease the learning curve considerably and would reduce my stress immeasurably. I live too far away from the "local" beekeeping organization for it to be useful in that regard.

Laurel Beardsley, Florida

Two different arrangements of boxes making up a beehive. Left: Two standard-depth boxes. Right: A similar-sized hive made up of three shallower boxes.

Supers

Supers are boxes of the same dimension as the original hive body that allow for nest expansion. Some beekeepers routinely call all boxes that make up a hive supers. Others prefer to call the main body a **brood chamber** and those stacked on top **honey supers**. The term "super" refers to the superior position of the additional boxes, typical of the vertical beekeeping style described in chapter 11, Additional Strategies.

Several depths of supers are available. In temperate climates, two full-depth, 9⁹⁄₁₆-inch-high supers (brood chamber and food chamber) often make up the basic hive. Other beekeepers, especially in warmer climates, manage single brood chambers only, restricting the queen with an excluder (see page 72).

Supers are usually placed on top of (in a "superior position" to) the hive body or brood chamber.

Additional supers can be the same depth as the brood chamber but are usually shallower, to reduce weight. A standard super full of honey can weigh 100 pounds; the shallowest super results in a box a little more than half that weight. Older and otherwise physically challenged beekeepers prefer shallower boxes.

Frames and Foundation

Ten combs fit in the standard hive body (brood chamber), each surrounded by a frame. The frame is designed so that the critical **bee space** of about 5⁄16 inch is left on all sides between it and the interior wall of the hive. The bees do not build into or glue up this space.

Frames can be wooden or plastic. The latter is becoming more acceptable to beekeepers, particularly those concerned about beeswax contamination from mite treatments: plastic frames are now readily available, some with integrated plastic foundation. Other combinations are plastic frames with wax foundation and wooden frames with plastic foundation. Each frame has a top bar, a bottom bar, and two side bars assembled into a rectangle that holds the comb.

Frames are fitted with sheets of **beeswax foundation**; aptly named, it is the template from which bees build their nest. Foundation is commercially produced out of beeswax or plastic and embossed with a pattern of hexagonal cells. The bees then draw out the walls of the cells based on the template (foundation). You can reinforce the foundation with embedded wires or retaining pins. The embossed cells on the foundation usually measure 5.1 millimeters wide; however, some beekeepers prefer to use smaller foundation in the 4.9-millimeter range.

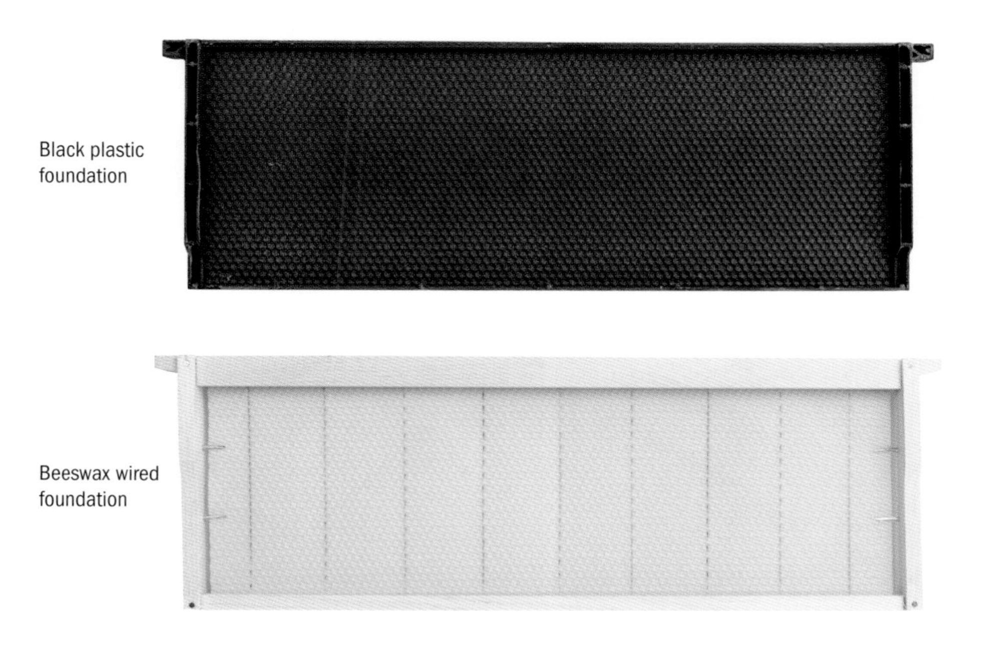

Frames can be of wood or plastic with beeswax or plastic foundation.

HOW A FRAME IS CONSTRUCTED

1. Remove wedge from top bar.

Despite all the options, a beekeeping newcomer should begin by assembling wooden frames, inserting wax foundation, and inserting retaining pins or wiring it for stability. Beeswax foundation inserted in wedge-style top bars and split-bottom bars are the standard.

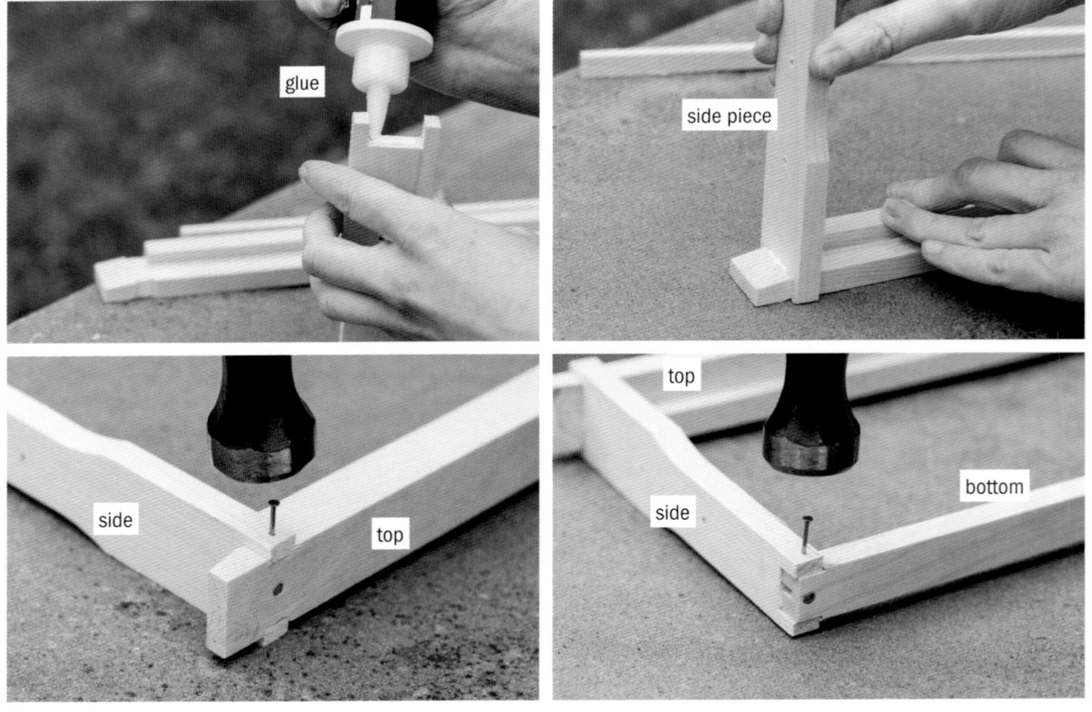

2. Assemble frames with glue and nails: each has a top bar, two side bars with retaining pin holes, and a bottom bar.

INSERTING FOUNDATION

3. Insert prewired foundation (shown here with embedded vertical wires with hooks). Ensure hooks are in the groove from where the wedge was taken.

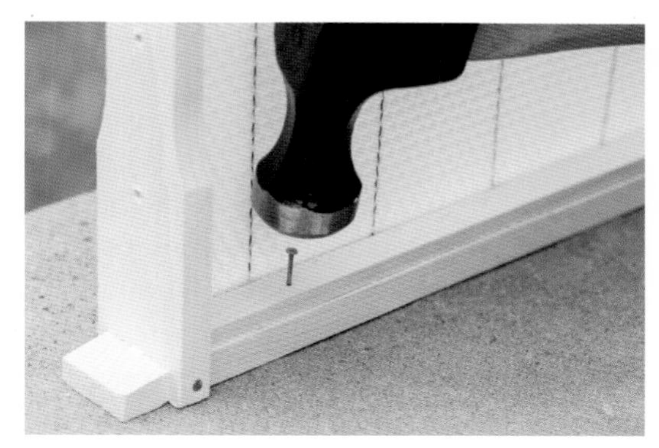

4. Replace wedge and nail it to top bar against the wire hooks.

5. Insert retaining pins in end bar to stabilize the foundation.

ABOUT FOUNDATION

Foundation is aptly named. It is patterned beeswax given to honey bees to guide them as they build comb. It also allows the insects to conserve energy because it provides a good deal of wax to make up the inner midrib of the comb.

Numerous kinds of foundation are available, depending on the beekeeper's need. Foundation used in the brood nest is thicker than foundation used strictly to produce honey comb. Bees prefer their foundation to be beeswax, but in recent years, plastic (coated and not coated with beeswax) has become available with some success. Foundation may be reinforced with vertical wires, which can produce extremely strong comb, useful in outfits that extract honey at high speeds.

A current debate about foundation focuses on the size of the cell that results from the template provided. In nature, hexagonal cells of the comb can vary, ranging from 5.2 mm to 4.9 mm. This results in bees of different sizes, which may affect the colony's behavior in certain ways not fully understood.

Historically, worker honey bees were thought to be more productive and healthy if they were produced in larger cells. Some believe this has developed a bee that is too big.

Enter those who think the time has come to reduce cell size because of reports that smaller bees from smaller foundation are less affected by varroa. So far, there's not much direct scientific evidence of this, but the "small cell" folks might be on to something. Time will tell.

Bottom Board

The bottom board, the floor of the hive, has a number of functions. Those available in supply houses are usually reversible, providing the option of a small entrance for winter and a larger one for the rest of the year. The bottom board is also critical for moving colonies.

You can modify the hive floor to fit other accessories such as feeders, pollen traps (for scraping pollen off bees' legs as they enter), or dead-bee traps (to measure colony mortality).

Recently, the bottom board has become a tool for monitoring and controlling the varroa mite. A greased paper or "sticky board" placed on the bottom board will collect mites that fall from the brood nest. You can then count the mites to determine the infestation level. The "open" or "screened" bottom board allows for mites that fall off either brood or adult bees to be permanently removed from the hive because once on the ground they cannot get back in. (See chapter 10 for more on managing varroa.)

Some beekeepers abandon the bottom board altogether and manage colonies without any floor. This is a relatively new idea for varroa control. At the moment, there are few standard recommendations with respect to whether bottom boards should be used for mite control. Once again, the novice should begin with the traditional configuration provided by supply houses.

Queen Excluder

The **queen excluder** is a barrier made of bars spaced so that worker bees can squeeze between them, but the larger queen cannot. Traditionally, the beekeeper places the

TWO KINDS OF BOTTOM BOARDS

The sliding tray allows the beekeeper to monitor varroa mite fall.

The screened board functions as a mite-monitoring (and possible mite-reduction) device because the mites fall to the ground and are thus removed from the colony.

excluder above the brood chamber to prevent the queen from entering the honey super and laying eggs. This keeps honey supers from being contaminated with brood, since most beekeepers prefer to harvest solid combs of honey.

Many beekeepers, however, call this device a "honey excluder," believing that workers with large nectar loads are also hindered in their efforts to move freely around the hive, thereby reducing final honey production. In some cases, this appears to be true, especially with a weak nectar flow compounded by a small worker population. Beginners, therefore, should use it sparingly, and only when strong nectar flows with optimal bee population warrant its use.

A queen excluder confines the queen to a specific area while letting workers in and out.

PROS AND CONS OF QUEEN EXCLUDERS

Few questions in beekeeping circles seem to provoke as much argument as whether to use queen excluders. Vocal proponents exist on both sides of the issue. A prevalent idea by those who condemn use of excluders is that engorged bees are bigger and so will not fit through the carefully measured slats of the excluder. The logic follows that a good deal of honey is thereby prevented from entering the super. Management procedures by these beekeepers apparently avoid problems that queen excluders are designed to eliminate, such as the presence of either brood or queen in honey supers removed for extraction.

It is well known that once a layer of ripening nectar is deposited above the brood nest, the queen will rarely cross it to lay eggs. Beekeepers who don't want to risk losing the queen or contending with brood in partially filled supers at extraction time, however, stick strictly to excluders. And they argue that excluders do not adversely affect honey production. Confining the queen to certain areas of the nest is also standard beekeeping practice in many queen-rearing operations. This would be impossible without excluder technology.

Who's right about excluder use? Both sides are. The decision whether to use the technology depends on its perceived usefulness in specific operations.

Covers

Most supply houses sell telescoping outer hive covers and ventilated inner hive covers. Both are standard products the novice should acquire. Many large-scale beekeepers opt for what is called a migratory cover, often nothing more than a piece of plywood used without an inner cover. These are cheaper than the telescoping variety, but not as secure. An advantage, however, is that they allow hives to be stacked together tightly when moved by truck.

The inner cover can be used in a number of ways. Most have a rim and an oblong hole in the center to accommodate a bee escape device. The rim is notched, offering the possibility of increased ventilation.

The telescoping outer cover protects the colony from weather, while the inner cover with its oblong hole can be used in a number of ways to separate boxes, ventilate, and feed colonies.

Hive Stand

In most areas, the hive stand is essential. It keeps wooden hives off the ground, protecting them from dampness, dry rot, termites, and other insects. Concrete blocks, rubber tires, and chemically preserved wooden rails might all be used to construct a stand. In areas where ants are a problem, the hive stand can be a rack perched on narrow legs that are inserted into cans of oil or water, which keep these insects at bay.

Hive Scale

The hive scale is a useful tool for the novice beekeeper to determine how a colony is doing without disturbing the bees. A hive that seems light in the middle of a major nectar flow, or when hives around it are noticeably heavier, is an immediate cause for concern.

Platform scales are traditionally used to weigh hives. Clever beekeepers, however, have invented many different kinds of devices based on bathroom and other kinds of scales, and have described and demonstrated them in the literature. In this age of computers, some beekeepers connect scales directly to electronic monitors. Weighing a hive is worth nothing, however, if the commitment to write down the data is not there. At least one national initiative to collect hive scale data has been implemented using the Internet by the Bee Informed Partnership (BIP). (See page 205.)

Feeders

The **Boardman feeder**, an inverted glass jar that fits into the entrance of the hive, is most commonly sold to novice beekeepers and is a good choice for beginners. A major advantage of the Boardman is that the syrup level can be monitored and the syrup supply replenished without dismantling the hive. There are several disadvantages to this feeder, however. It doesn't hold much food, and if used with weak colonies, it may incite robbing behavior.

Other types of feeding arrangements include modified frames (**division board feeders**) and **top feeders**, some of which can be quite elaborate. Plastic bags filled with syrup can be punctured with small holes and placed over the frames.

Pollen Trap

In some cases, a hive entrance can be modified as a pollen trap. As the bees enter the colony, pollen is scraped off their legs and drops into a container that later can be collected by the beekeeper. Trapped pollen can be either fed back to bees or sold as human health food.

Tools of the Trade

A couple of items should go with you every time you visit the hives: a smoker to reduce the bees' defensiveness and the all-purpose hive tool.

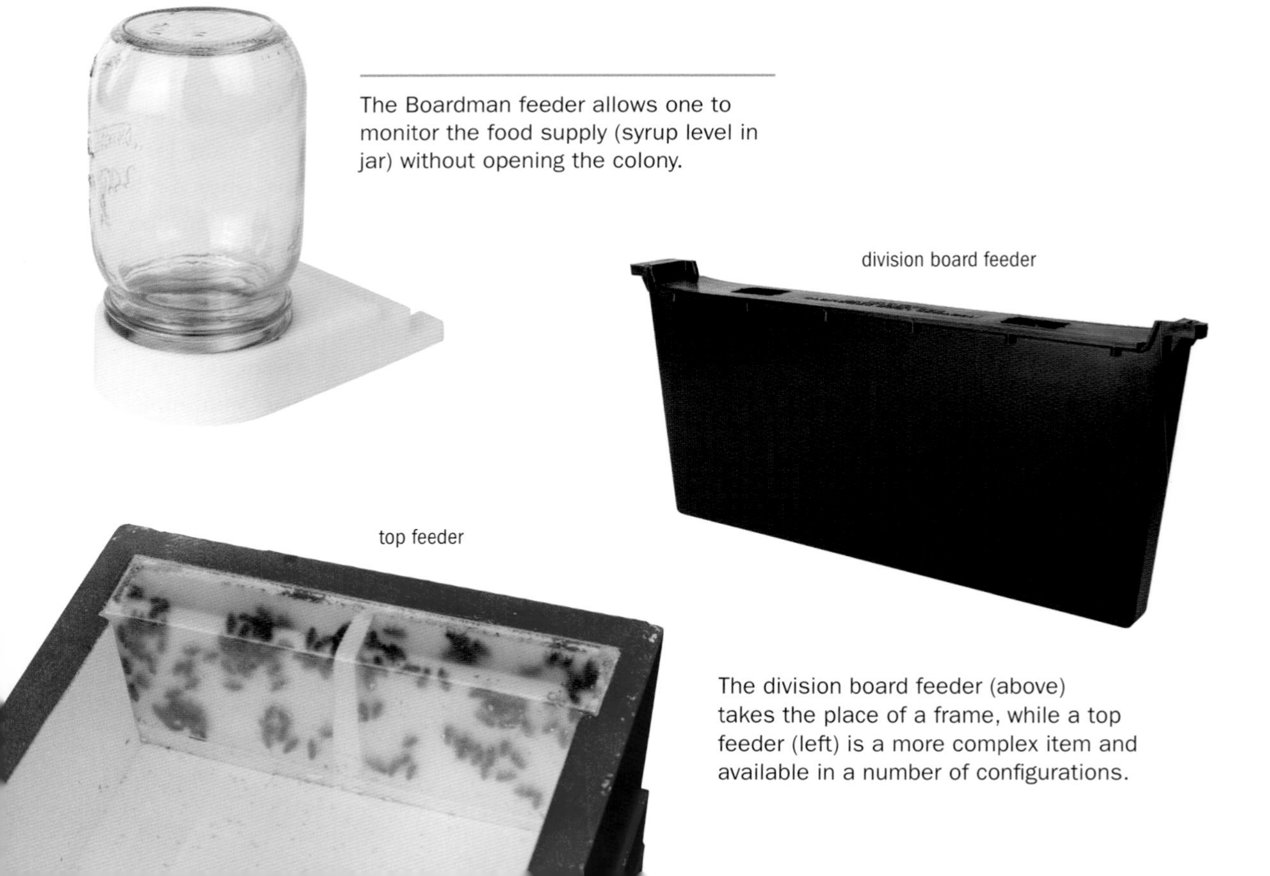

The Boardman feeder allows one to monitor the food supply (syrup level in jar) without opening the colony.

The division board feeder (above) takes the place of a frame, while a top feeder (left) is a more complex item and available in a number of configurations.

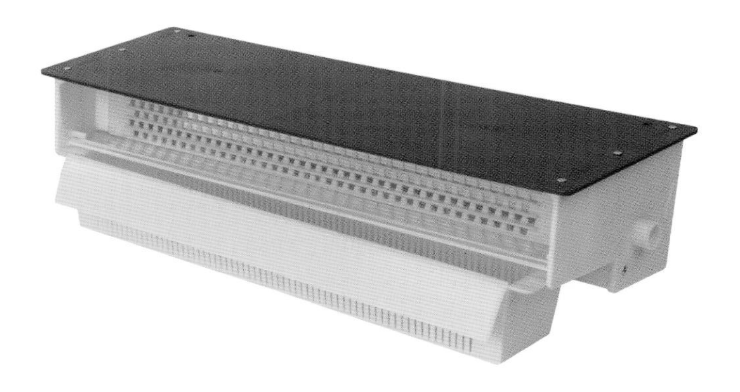

With a pollen trap, the normal hive entrance is blocked off, and an entrance above it allows pollen to be collected from foraging bees in a front-mounted box.

The Smoker

The smoker is a most important and essential beekeeping tool, although you may find some beekeepers who claim not to need it. Smoke suppresses defensive behavior, and you should have a smoker lit any time you are in the beeyard.

Smokers usually come in two sizes, 4 × 7 inches and 4 × 10 inches. (Beginners should get a 4 × 7.) Stainless steel models will last longer. A shield to prevent direct contact of the hot barrel with the beekeeper's skin and clothes is most desirable. (For more on using the smoker, see chapters 6 and 7.)

Essential beekeeping tools include the smoker (above) and the hive tool (below). The latter can be various shapes and sizes; a standard one is shown here.

Hive Tool

Called the "universal tool" by some, the traditional steel hive tool is hard to beat for the variety of jobs it can do. Its main job is to remove frames from colonies and to scrape debris off hive parts. Several shapes have been devised, and you can paint yours in bright colors so it can be easily seen when you leave it behind in the field. In a pinch, other things like screwdrivers can be used to manipulate colonies, but they can damage your equipment. Beginners should buy two tools, one for each hive they start out with as recommended in this book.

Beekeeper Garb

There is a wide range of protective clothing available to beekeepers. It is always wise to wear more when beginning a hive manipulation because it is usually easier to take off than put on beekeeper garb. Depending on your reactions to stings, you may decide whether to be fully protected. Only one item is essential, however: you can remove any piece while inspecting bees except the veil. Although you can recover from the effects of stings no matter where they might occur on the body, should you get stung on the eyeball, severe damage would be done, perhaps resulting in blindness. Never manipulate bees without a veil.

A full bee suit with veil and gloves provides optimal protection.

Veil and Hat

Novice beekeepers can purchase a variety of head gear to support that critical accessory, the veil. Wear a veil at all times to protect your face and head from stings where they are most dangerous. Veils are designed to be worn with helmets and hats, or in some cases without a hat (the Alexander model). The square folding type is more durable over time because you can easily pack it away when you are not using it.

Gloves and Coveralls

Gloves give the novice confidence. They are often abandoned once experience is gained, but can become necessary in an emergency. Several kinds of gloves are available, including rubber ones, but the traditional canvas or leather glove with a long cuff designed for beekeeping is best for beginners. Coveralls range from all cotton to a space-age model made with breathable foam.

Helmeted veil

Hooded veil, which can zip to a bee jacket

Nitrile gloves, affording more dexterity but less protection

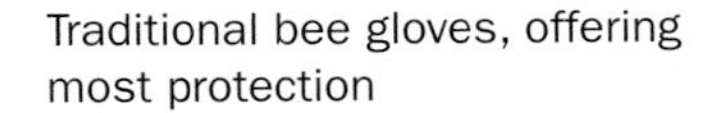

Traditional bee gloves, offering most protection

6

ENTER THE BEES

There are five common ways to obtain honey bees: in a package, as a nucleus hive (**nuc**), as an established colony, from a swarm, and from a feral colony. All of these will be discussed here, but novices should consider only the first two. Outside of beekeeping circles, it is hard to find good sources of honey bees, and that's another reason to subscribe to specialized beekeeping publications and join local and national associations.

No matter how you acquire your bees, you must first become acquainted with the smoker, that all-important beekeeper's tool, so we will begin with it.

Smoker Basics

The smoker is a primary tool for controlling bees while you work. Use of the smoker requires experience and patience and is an art form itself.

Fuel

Practically anything that burns is good smoker fuel, from pine needles to dried dung. Avoid any material that has been chemically treated, such as coated burlap or pressure-treated wood. Chop seasoned ½- to ¾-inch-thick limbs into 3- to 4-inch lengths. Seasoned chips or chunks from logs or stumps also work well.

Lighting the Smoker

1. **Have your fuel handy.**
2. **Crumple up one-half to three-quarters of a sheet of newspaper** so it will fit into the smoker, but don't wad it so tightly that it won't burn easily.
3. **Light one end of the paper,** put it into the smoker, and start puffing the bellows. At the same time, use your hive tool to push the paper down into the smoker and get it burning well.
4. **Add some small chips on top of the paper.** When they are burning nicely, add bigger chips until you have filled the smoker. The smoke produced should be cool to the touch. Guard against hot smoke or flames. To cool the smoke, pack some green material like grass clippings on top of the burning mass in the canister.
5. **Close the lid.**

This apiary is in perfect position as the adjacent orchard begins to bloom.

If the fire goes out at any point in the process, shake everything out of the smoker and start again. Don't get frustrated—it may take three or four tries.

After half an hour, about half of the fuel will be burned and you should fill the smoker again. When you have finished for the day, plug the snout of the smoker with a wad of paper or grass clippings or a cork. The fire will go out and preserve the half-burned fuel to make the next lighting easier.

Using the Smoker

1. **Put a hook on the bellows of the smoker** so you can hang it from your belt while you move around, or hang it from the end of the opened hive so it will be handy while you're working with the bees.
2. **Give the hive a couple of puffs of dense, white smoke** in the entrance when you approach. If you are standing between two closely spaced hives, give both a little smoke in the entrances.
3. **Use your hive tool to pry up the cover.** Immediately waft smoke over the frames and gently direct it into the colony.
4. **Replace the cover** and wait at least two minutes before removing the cover again. (See The Tao of Smoking Bees, page 103.)

Package Bees

Novices should begin with package bees. In the United States, most packages originate in the South and the West so they can be ready for the needs of more northerly beekeepers in early spring.

A beginner can more easily handle a package than a nuc because the bees are not a true, organized colony, but just a bunch of insects in a box. They lack a home to defend, have limited organization, and little sense of purpose. They are, therefore, among the least defensive of honey bees.

Bees in a package have been stressed in many ways. They have been separated from their original queen, endured the rigors of a trip through the U.S. mail or other delivery method, and perhaps suffered other indignities such as being left out in the rain or direct sunshine. Because of this, package bees must be transferred into a hive without delay.

We recommend you start by establishing two colonies as a hedge against complete failure, should one colony not be successful. It can be extremely satisfying to watch your package develop into a fully functioning honey bee colony. But it is a given that few, if any, colonies established from package bees will produce a surplus honey crop the first season — except in the far North where, thanks to the long summer days, they can readily produce a 300-pound crop.

Package bees clustered in a screened box ready for installation.

WHAT IS A PACKAGE?

A package is a screened cage packed with honey bees. The unit is really an artificial swarm, taken from a large colony and shaken through a funnel into a screened box. A caged queen comes included in the package, along with a can of sugar syrup to sustain the bees on their journey. Blocking the entrance of the queen cage is a candy plug, a sugar candy that the worker bees will ultimately eat to release her into the hive.

The number of bees in the package is determined by weight, the most common size being 3 pounds. With a package holding about 3,500 bees to a pound, the beginner typically receives about 11,000 bees to install. Producers generally take care to ensure that the majority of bees in a package are workers, but drones may be seen on occasion.

In the past, packages could be expected to be free of diseases and pests. This is no longer the case. At least some parasitic mites are probably found in any package, and some packages have more than others. Nevertheless, the unit is still not as potentially problematic as a nucleus or established colony, both of which come with comb, which can carry mites. Varroa mite management should not become an issue until the colony develops a sizable population.

Installing Package Bees

Any brand-new hive that receives a package is considered a fragile work in progress. Remember that the bees have been treated roughly and will have a tough time as they transition from a purposeless mass of insects in a screened box to a productive colony in a beehive. They need a lot of tender loving care as you introduce a new queen and establish a feeding regimen. Here are some tips and guidelines.

Storing and Feeding a Package

Install a package as soon as possible after it arrives. Any delay will result in fewer live bees in the colony. That said, packages are designed to be stored for a limited time. A maximum of a week might be possible, but every day of delay is critical. It is important, therefore, that the beekeeper be ready to make a rapid installation.

If a package needs to be stored, feed the bees by dripping sugar syrup (one part sugar mixed with one part water, by volume) through the screen. Add just enough syrup to dampen the bees. Turning the package on its side makes the feeding process easier. Some package designs have an enclosed bottom to hold syrup on a temporary basis.

Do not feed by brushing syrup on the screen with a paint brush, as some recommend. The danger of damaging the honey bees' delicate mouth parts and limbs with the brush is too great.

Gently drip syrup on the screen to feed the bees if storing before installation.

BEEKEEPER'S STORY

I WAS FIRST INTRODUCED to beekeeping when I was about 8 years old, almost 50 years ago (I can't believe it). Opening a beehive for the first time was like being a guest visiting a special place.

The experience remains as clear in my mind as if it happened yesterday.

The school class visited a local beekeeper who had two colonies in his backyard. It was a warm sunny afternoon, and when the cover was removed, I was totally overwhelmed seeing these thousands of bees, the soft buzzing sound of contentment, the colors, the fragrance of warm wax and honey: it was magical! It didn't take long for me to get my own bees. What attracted me to beekeeping more than anything was the opportunity to witness and experience the inextricable link of the bees and flowering plants; the idea that all living things are interlinked, connected, and therefore interdependent. Small children have the ability to observe nature intently and completely, but as we get older we seem to lose this ability. For me, being involved with bees gave me the opportunity to retain this closeness to the natural world around me.

I never owned more than four colonies at any one time. Later, in Canada, I became a research assistant and worked on hundreds of colonies owned by a commercial beekeeper. I also headed a beekeeping development project in Uganda for CARE International, after which I assumed the position of Provincial Apiculturist of Alberta and then was appointed Provincial Apiculturist of British Columbia.

Beekeeping has become a far more complex enterprise today than 30 years ago, when there were few diseases and fewer demands placed on the colony as a production unit. We beekeepers must continually relearn the basics of good management practices. Instead of always thinking about what we can take out of the hive, we must think what we can put into the hive so that the colony will prosper in health and productivity.

A far greater emphasis, therefore, must be placed on training and education. Only educated beekeepers will be successful in practicing Integrated Pest Management techniques that are needed to retain a viable future honey bee resource, and to produce hive products of the highest quality in a sustainable and environmentally friendly manner.

Paul van Westendorp, British Columbia, Canada

Installation

If possible, choose a calm, sunny day to install the package. It is best to set up a colony in the morning (9:00 a.m. to 11:00 a.m.), before the bees become too active. Read the following steps several times before proceeding so that you have the sequence clear in your mind.

1. **Make sure everything is laid out** and ready for the installation. Tools needed include a hive tool, a pair of pliers for removing the food can, and a knife or thin nail for puncturing the candy plug in the queen cage. You also need your veil and gloves, a smoker, and the feeder.

2. **Fill the top feeder** with sugar syrup and have it ready.
3. **Remove three or four frames** from the center of the hive and lean them against the hive body.
4. **Install a standard entrance reducer** if a top feeder is used; with a Boardman or entrance feeder, use grass or leaves to obstruct the opening so that only a few bees a time can leave or enter the hive. Do not close the entrance entirely.
5. **Put on a veil and light a smoker** in the unlikely case it is needed. Remember, package bees, like swarms, are not usually defensive, so coveralls and gloves are not generally necessary, but a veil is.
6. **Using the hive tool, gently pry the lid off the package.** Bees should not escape, but even if they do this is not a problem. They will return to the colony in due time.

To install package bees you need a hive tool, a pair of pliers, and a thin nail. Be sure to equip yourself with veil, gloves, and smoker **(step 1)**.

Fill the feeder with sugar syrup **(step 2)**.

Remove frames from the center of the hive to create a space to receive the bees **(step 3)**.

Install an entrance reducer, as shown, or reduce the size of the entrance with grass or leaves **(step 4)**.

Remove the queen cage from the package **(step 7)**.

7. **Remove the sugar syrup can and the queen cage.** Cover the holes with the package lid to prevent bees from exiting. Again, ignore any bees that take flight; concentrate on the job at hand.

8. **Examine the queen** and make sure she's alive. A colony established with a dead queen will lose many workers and may abandon the hive. If she is dead, call the producer and arrange to have another sent immediately. The package can usually be stored for a few days as noted on page 84. Some queens are caged with worker attendants that feed and groom her. If you can, remove the attendants without letting the queen escape.

Remove the feeder can with pliers **(step 7)**.

Check that the queen is alive in the cage **(step 8)**.

9. **Remove the cork** or rotate the metal disk to expose the candy in the end of the queen cage. Make a hole in the candy with a thin nail, being careful not to injure the queen inside.

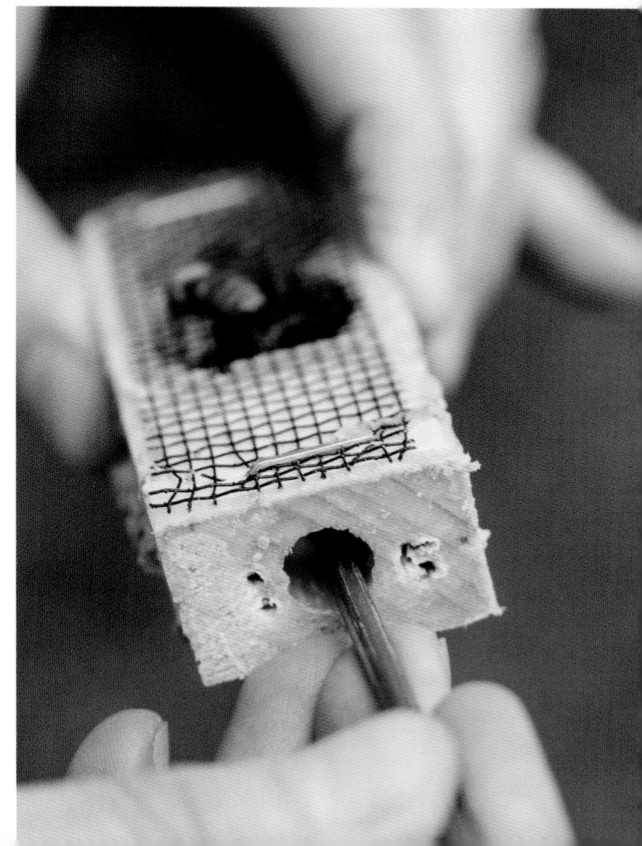

Perforate the candy plug with the thin nail before installing the queen cage **(step 9)**.

Dump the bees into the vacant spot in the center **(step 10)**.

10. **In one smooth motion, invert the package and dump the bees into the space in the hive body you created by removing the frames.** Don't worry about getting them all in; only a critical mass is needed to attract the outliers.

11. **Replace the frames** in the hive gently so as not to damage the bees.

12. **Install the queen cage** by wedging it, candy down or up, between frames near the inside wall of the hive body. Try to ensure that the screen on the cage is not totally blocked so the bees can feed the queen through it. This can be challenging, so try not to obsess too much about the orientation of the queen cage. Indirect introduction of the queen allows the bees to become organized and free her slowly over time, reducing the risk of her being lost or killed in the initial confusion.

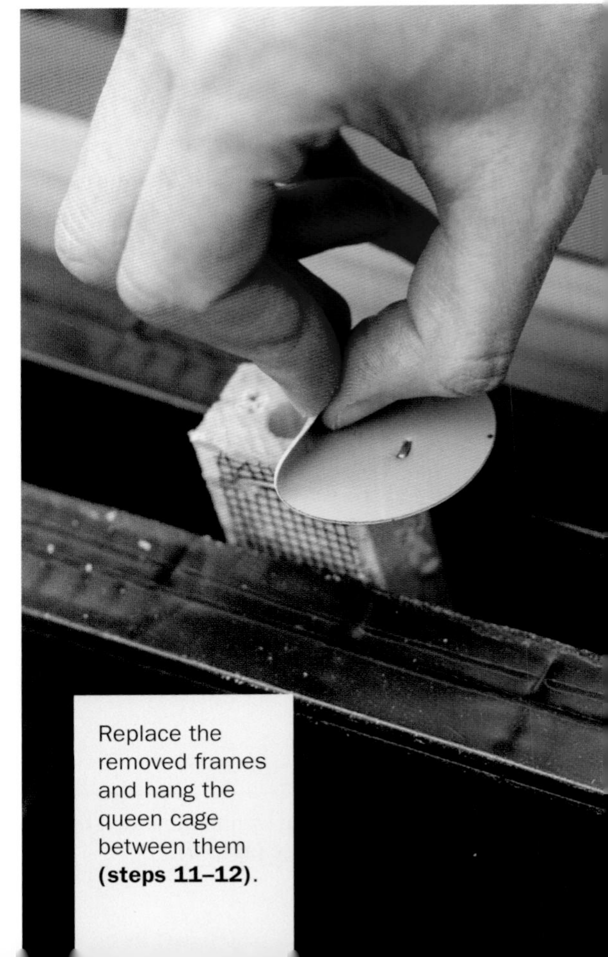
Replace the removed frames and hang the queen cage between them **(steps 11–12)**.

Place the opened package in front of the colony **(step 13)**.

13. **Put the empty package near the entrance.** Bees that are still inside it or flying will enter the colony on their own.

14. **Install the feeder.**

Install the top feeder on the hive. There are several styles of feeders; many beginners use a Boardman feeder, which is installed in the entrance rather than on top of the hive **(step 14)**.

After They Are Installed

The next day, make sure that the entrance to the colony has an opening for the bees to enter and exit; do not remove the grass plug entirely, however, so the small population can effectively guard the entrance.

Leave the unit alone for at least a week to allow the colony to release the queen by eating through the candy plug. When a week has passed, check only on the queen's status by briefly opening the hive.

Continue to fill the feeder as it becomes emptied. Package bees must be continuously fed until they stop taking syrup, which can take as long as eight weeks. Observe the syrup levels in the feeder carefully to determine when you can stop replenishing it. Err on the side of giving too much rather than too little.

USING WATER TO INSTALL BEES

Some experienced beekeepers dip the queen cage in water, wetting her wings so she cannot fly, remove the screen, and introduce her directly into the mass of bees. The bees are probably used to this queen after their journey and consider her their own. This can reduce the time for the package to become an established colony. Others shake water through the screen of the package to moisten the other bees and reduce flying. The insects clump together when wet and can be more easily dumped as a mass into the hive.

Managing Package Bees

Managing package bees is somewhat different from managing established colonies. Because it is a fragile unit, an installed package requires more attention, and beekeepers must carefully monitor progress and assist where needed.

Remember that the population is small, about 11,000 bees to start. Workers live only four to five weeks; it takes three weeks for a functioning adult to develop from an egg and maybe two more before foraging can begin. Initially a colony's population will, therefore, rapidly decline.

If established on beeswax foundation, the comb must be drawn (built out from the embossed base). Bees recruited by the colony for this task are not available for other duties. The new queen should begin to lay eggs as soon as possible to get a suitable brood nest going. Nectar and pollen may be scarce. The most important thing the beekeeper must be concerned with is food supply. The mantra should be feed, feed, feed!

First Inspection

The first inspection, a week after installation, is important and should be brief. There is one objective: to see whether the queen has been released and is functioning.

1. **On a sunny day,** light the smoker and don protective gear.
2. **Observe the activity** around the hive before attempting to remove the cover. Are bees actively entering and exiting? Do they have pollen pellets on their legs? Does the activity look normal and relaxed or do the bees appear agitated or excited? Take all this in, but again, remember the primary objective of the visit.
3. **Check the smoker** to ensure the smoke is cool, not hot. The smoke gets hot when there is too little material in the smoker. Add a little fresh green grass to the smoker to cool the smoke if necessary.
4. **Puff several times,** sending smoke into the entrance; remove the cover and again puff several times into the top of the hive.
5. **Replace the cover,** wait at least two minutes, and remove it again.
6. **Locate the queen cage.** It should be empty. Take it out and put it aside for the time being.
7. **If the bees have built comb around the cage,** attaching it to the foundation, scrape it off with your hive tool as it will interfere with subsequent construction.
8. **Remove an outside frame** and prop it outside against the hive to provide room to work the rest.

9. **Slowly take out each frame,** one at a time, examine it, and replace it. Try to spot the queen. Look for eggs and larvae in the cells.

10. **When you have finished your inspection,** and either found the queen or determined that she is functional, replace the combs and close up the hive.

FIRST INSPECTION

Light the smoker **(step 1)**.

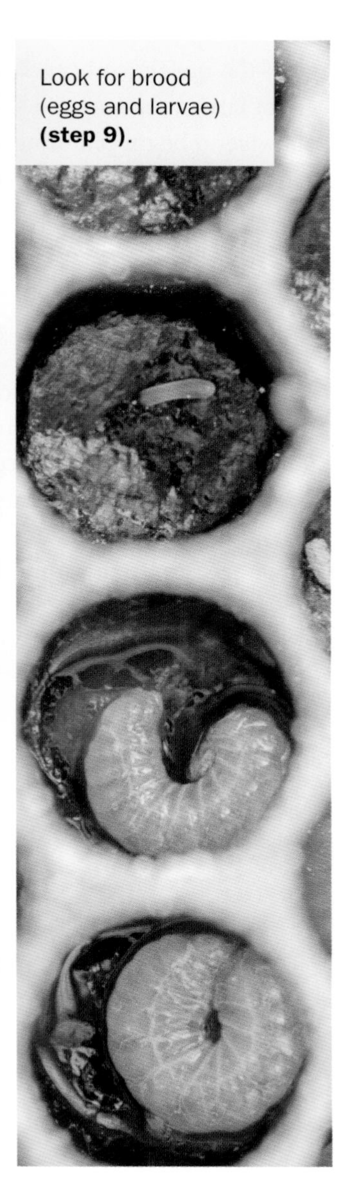

Look for brood (eggs and larvae) **(step 9)**.

Remove queen cage to see if she is released **(step 6)**.

It is not necessary to actually see the queen during this visit. What you are really checking is the state of the colony as a whole. If, therefore, when examining the frames, you see that cells are being drawn and eggs or young larvae in jelly are present, that is sufficient evidence that the queen is on the job. Replace the combs and close up the hive.

If the queen is not seen and there are no eggs or larvae, she may still be functioning, but it is wise to begin planning to get a replacement. Go back into the colony a day or two later. If there is still no sign of brood, assume the colony is queenless and acquire another as soon as possible.

Ongoing Monitoring

Assuming signs of a functioning queen are present (in the form of eggs and larvae), begin to monitor the colony on a weekly basis to view its progress. Again, these visits should be brief, only sufficient to determine the population's characteristics, such as number of frames of food and brood, absence of disease, brood-to-adult ratio, and food supply.

It cannot be stressed enough that it is vital to continue feeding for at least three weeks, stopping only when the bees no longer consume the syrup. As time goes on, the colony should begin to fill with bees. At that point, it can be considered established and can be managed like any other colony in a beekeeper's operation.

Installing a Nucleus Colony

A **nucleus colony** (nuc) is a small but complete colony containing brood, adults, and a laying queen. It may comprise three to five frames and be delivered in a throw-away or returnable box. Many beekeepers have success with nucs. They are widely used to increase colony numbers.

As a functioning colony, a nucleus is much less fragile than a package. Because it is small, the residents are not generally defensive, but the beekeeper must be on guard because the

Nucleus colonies (nucs), delivered by a supplier, with a specialized entrance that can be sealed off completely during transit.

unit is much better organized than any package colony. It also comes complete with combs, which increase the risk of disease introduction. On the other hand, the frames should contain stored food, which is an advantage over the use of package bees that require continuous feeding.

Some suppliers install nucs when the buyer brings a standard hive body to the seller's establishment. Otherwise, if the nuc is delivered in a reusable container, the frames must be relocated to the center of a waiting brood chamber. Be sure to have a smoker lit and handy just in case.

Lightly smoke the bees and transfer all the nuc frames into the center of the hive body. There should now be a small, functioning colony in the middle, flanked by five to seven frames of foundation or drawn comb. Install a feeder for the colony and put on the cover.

Because it already has brood, the nucleus colony will probably contain a population of varroa mites. It should be a small population if the seller has done a good job managing them, but this should not be taken for granted. Begin to monitor the mite population immediately to determine its potential to adversely affect the nucleus's development (see pages 172–174).

Starting with an Established Colony

Purchasing established colonies is risky for the beginner. A fully functioning unit can be defensive, and the beginner may simply not be up to the task of manipulating it. It also doesn't provide the learning experiences that can be had when assembling a box and frames with foundation, and when installing package bees. Another disadvantage is that the history of the unit may not be known. It might have been neglected and be substandard; novices will not have the experience necessary to detect deficiencies such as brood diseases, queenlessness, and high levels of varroa or tracheal mites.

Varroa Mite Alert

One of the biggest risks involved in acquiring an established colony is the potential for also acquiring a devastating varroa mite population. This will affect not only that specific hive but others in the area as well. The previous owner should have some record of mite management in the past. If not, the mite population must be determined before you make any plans to move the colony to a new location.

One way to minimize the risk of varroa or other deficiencies is to have the unit examined by an experienced beekeeper from a local association or a state inspector before purchasing it. In some states, the law requires beekeepers to be registered.

Swarm!

Many beekeepers have begun their careers by catching swarms. Free bees can be a great incentive to begin this adventure. Unfortunately, there are fewer swarms than there once were because of increased diseases, varroa mites, and habitat destruction.

The window for collecting and installing swarms successfully is also small. In the Midwest, the saying still goes that "a swarm of bees in May is worth a bale of hay, but one in July isn't worth a fly." Swarms collected later in summer are far less likely to succeed, because the scarcity of nectar-producing plants makes it difficult for the colony to build stores and population sufficient to ensure a successful first winter. Generally, the beginner should stick with an artificial swarm in the form of package bees.

If you have to act quickly and lack the proper equipment, you can pick up a swarm in almost any kind of container of suitable size, including a cardboard box or pail. It is helpful to place a stick in the container for the bees to hang on. You must be able to close the container tightly so the bees can't get out — but without shutting off the air supply. Most use some kind of metal screening for this purpose. If there is not adequate ventilation while they are transported, the bees will die, leaving a huge mess.

Beekeepers should always have equipment ready to pick up a swarm. An old hive body with plywood nailed to the bottom, and a screened top, is a good option. Frames with foundation in such a box will make a fine nucleus for a new colony.

Although swarms of bees do not usually sting, because they are disoriented and not protecting an established nest, they can be defensive in rare instances. Of particular concern is a "hungry" or "dry" swarm that has consumed its food supply. There's no way to know this in advance, so beekeepers attempting to collect any swarm should always be prepared in case a situation gets out of control; they must be equipped with a smoker and protective clothing — and an unobstructed escape route!

In regions where Africanized honey bees are established, collecting swarming bees is discouraged, if not illegal. Some beekeepers collect them anyway, but at great risk to themselves and others. Many a beekeeper in the tropics has been fooled by these overly defensive bees.

CAPTURING A SWARM

A classic swarm clusters temporarily while scouts find a new nest.

Shake the swarm off the tree into a container.

STOPPING BEES

There may be certain situations when individuals wish to immediately eliminate an exposed group of bees. A swarm hanging in an unappreciative homeowner's yard or somewhere on a school grounds might be an example, as is an overturned truck load of beehives. The question arises as to what chemical can be sprayed to stop the bees and get them immediately under control, that is, totally immobilized and/or dead.

The answer in many cases is soapy water, a detergent solution formulated with 1 cup of dish-washing detergent in a gallon of water, applied using various sprayers. Because detergent acts as a wetting agent, the result is less surface tension present in the solution being sprayed. Instead of beading and running off, it wets the bees. This also prevents flying because the wings of the insect cannot get up to speed while soaked with water. Finally, the insects' spiracles, or breathing holes designed to ward off or repel water, are filled with the "wetter" detergent-water mixture. The bees can no longer breathe and suffocate. The technique is as effective as many others, including flame throwers. More importantly, it is safer for both applicator and the environment than almost anything else and requires no special label. It should not be overlooked by regulators and others who, under certain conditions, want to get an otherwise difficult honey bee situation under control.

The cluster, much reduced. Brush as many stragglers as you can into the container.

Dump the cluster of bees from the bucket into a box with a lid (such as this handy leftover "nuc" box, or a hive body on a bottom board), place the lid on top, and close the entrance with grass.

BEEKEEPER'S STORY

I LIVE NEAR A VILLAGE in the Chiltern Hills about 35 miles northwest of London, England, as the bee flies. We are 660 feet above sea level and our weather is slightly cooler than down the hill. Spring seems to appear a couple of weeks later than it does in London (at sea level). Climate? We don't have one; we only have bits of other peoples' climates. It's very variable. For example, at the beginning of June we had very warm 85°F (29.4°C) weather. After a couple of weeks, it changed to cool, showery, and breezy stuff: it's still raining and we're still waiting for summer.

The bee season here starts with cleaning floors on a warm day in mid-March. First inspection happens in April, and then we shut them down, with feeding done by the end of the first week of September and mouse guards on by the last week in September.

This is basically an agricultural area with small fields (wheat, oats, barley, oilseed rape, and a bit of flax), big houses with large gardens, and everything between. Usually, dandelion starts the thing off, then come chestnuts, and then oilseed rape in mid-April. After that, there are flowers in the gardens, then the "June gap," as it's called, then summer flowers with chrysanthemums and dahlias bringing up the rear, and finally, ivy in September.

Varroa and its resistance to chemical treatment is our biggest challenge. Pesticides are troubling, too, as they haven't been properly evaluated here yet. Lack of government funding for bee research is a problem as well. Our bee losses around here last year were knocking on 40 percent, this year 30 percent, and the worrying thing is that no one knows why. We better find out before it's too late. There is a lot of publicity about bees on TV and radio, as well as in the press. Stories are usually titled "Sting in the Tale" or "The Buzz about Bees," and this publicity has caused a massive increase in people wanting to start beekeeping.

I'm on the local council/police/fire brigade list as a "swarm collector." This year, I have had a few calls, mainly for bumblebees and, lately, wasps. Seems here we need a dictionary for the public to use. Anything above three insects is a "huge swarm," and anything about 6 feet above the ground means "You'll need a double extension ladder!"

Peter Smith, Buckinghamshire, United Kingdom

A swarm of bees fills the sky in search of a new home.

Management after Capture

Once a swarm is installed into a hive, and subsequently transformed into an established colony, its potential defensiveness greatly increases. Attempting to manipulate an organized colony in the same manner as the swarm it came from is a recipe for disaster.

The best defense against a rash of negative public relations is for the beekeeper to ensure that only gentle European bees are in an apiary. In Africanized honey bee country, which currently includes south Florida, south Texas, and west to southern California, many have adopted the philosophy "Just say no to Africanized honey bees." The consequence of this means giving up collecting any swarm from an unknown source.

A Wild or Feral Colony

Of all ways to obtain bees, the least desirable for the beginner is to attempt to move a wild or feral established nest from its location to a beehive. Honey bees are very good at hiding their nests; only an expert can determine its true size and location.

This is especially true if a colony is found in a building. Anyone attempting to remove a colony from a structure should probably have a contractor's and a pest control operator's license, especially in areas where Africanized honey bees are established. When in doubt, call a licensed pest control operator with the appropriate experience.

COLLECTING SWARMS AND FERAL NESTS

Swarms and feral nests of honey bees have always been a major resource for beekeepers. In certain areas, there is fierce competition among individuals interested in collecting these free bees. Traditionally, beekeepers have, therefore, initiated contact with fire and police departments, Extension offices, and other public agencies to be called when swarming bees are reported.

Although wild swarms are easily captured, they can be vectors for parasites or pathogens. It also now appears that European bees, especially in the southern United States, are subject to physical takeover by wild Africanized queens and associated swarms. More insidious is gradual Africanization of a population because of numerical superiority of feral Africanized drones available to mate with queens from European colonies. For all the above reasons, free bees could increasingly become a liability.

The following suggestions may minimize anticipated problems from feral nests and swarms:

1. Locate feral colonies and evaluate their potential for nuisance.
2. Locate potential future nesting sites for feral colonies and keep them under observation.
3. Learn how to kill feral bees and destroy their nests. Be prepared to do so when they are infested with mites or could potentially become a stinging hazard.

In summary, aggressive elimination of honey bee swarms and nests of unknown origin may be an important management strategy for beekeepers in the future. Potential benefits include: eliminating bees that might cause highly publicized stinging incidents, reducing competition of nectar collection on managed colonies, and eliminating sources of apiary Africanization and infectious diseases.

THORNE

7

MANAGING HONEY BEE COLONIES

The idea of even approaching a beehive, let alone opening it, is daunting. Most people are understandably tentative in the first few inspections of a colony. One way to make it a little easier is to find a mentor to help you get started. In no time, the butterflies in the stomach will be gone. Instead, you will become immersed in the colony's activities, eagerly looking at the different colors of bees, assessing the stores of honey and pollen, and finding a queen going happily from cell to cell depositing shiny new eggs.

Working a Colony

Before manipulating a colony you should keep some things in mind. The better the weather, the more pleasant the experience will be. Honey bees begin to cluster at about 57°F (14°C). It is not advisable to manipulate a colony when a cluster has formed. This disorganizes the bees and they may not have enough time to reorganize before the temperature drops again, which can lead to life-threatening complications. If a colony must be manipulated during cold weather, make sure it's early in the day.

Bright, sunny, warm days are much better than cloudy days for inspecting colonies. Bees can be highly defensive in the rain and on hot, muggy days, especially if there are thunderclouds. During these periods, workers may stop foraging, and this creates a "perfect storm"—too many bees in a colony with nothing to do except deal with the beekeeper. It's also thought that static electricity from these unstable weather conditions makes the insects irritable.

TIPS FOR WORKING A COLONY

- Choose a bright, warm, calm day, ideally during a nectar flow when field bees are absent from the hive
- Dress appropriately (see chapter 5)
- Inspect when the hive is not in shade
- Work slowly, calmly, smoothly
- Use a cool smoke and don't over-smoke the hive (see The Tao of Smoking Bees on page 103)

Proper use of a smoker is a key skill. A beekeeper should apply smoke deliberately and observe its effect, which can differ from colony to colony.

BEEKEEPER'S STORY

MY FIRST YEAR IN COLLEGE [I met] a beekeeper [who] had worked at the A. I. Root Company in Medina and appeared in several of its advertisements and brochures. He shared information about bees and beekeeping with me. It stirred my interest, and I had the benefit of spending time with him for a couple of years.

In early August after my first year in college, a swarm landed in my parents' yard. The local bee inspector convinced me they wouldn't survive, so I didn't capture them, but that also ignited my interest.

During my first year of teaching eighth grade, I mentioned something about bees. After class, a girl approached me and said her father, who was in the Air Force, was being transferred to Alaska that Saturday and had beehives he needed to sell. This was on Monday, and by Saturday I was the proud owner of nine hives of bees. That was 45 years ago. Everything skyrocketed from there.

Bees in rural areas don't produce as much as those in urban and suburban areas. During the early years, I kept accurate records and averaged 30 pounds of honey per rural colony, counting everything that had bees in it. I maintained one hive in my parents' yard in a suburban housing development, and it was a rare year that it did not produce 100 pounds of surplus honey. I'm not sure I can tell how much nectar production has shifted in terms of plant species used, but where the honey is produced is clearly in the areas where people live.

This year we had a wonderful locust honey flow; that happens here only about once every 10 or 12 years. There is lots of clover in the yards that the bees work heavily, but I don't know how much nectar they get from it. Along the roadsides, yellow and white sweet clover grows, and I have wondered how much it produces. The smell of goldenrod in the apiary in fall tells me the bees collect significant amounts.

I am running about 30 hives now, and I am in the process of getting bees into all the empty equipment I have accumulated through the years, rather than letting it stay in storage. Clutter is such a challenge for me as a beekeeper!

Beekeepers around here primarily use two deeps, but a lot of beekeepers are going to three mediums to make lifting easier. I also am finding more beekeepers getting eight-frame equipment now, especially since it seems to be easier to obtain from suppliers.

I detest spending money to buy bees. For the most part, I split hives and use swarm cells as well as queens I graft myself. I do buy queens when I don't have any available, but I resist paying $20 for a bug when I can raise them myself. Because it takes a significant investment of time to raise queens properly, I don't do it as much as I should. I use Italian bees primarily.

Bill Starrett, Ohio

No doubt, the best time to work bees is during a nectar flow. You can tell when a nectar flow is in progress by examining the frames for fresh nectar: it will be so liquid that it can be shaken out of the cells. You may also notice increased activity in nearby fields and bees on open blooms, but these observations are less reliable than looking at the combs and seeing fresh product.

Of course, it is not always possible to work bees during nectar flow. And if the flow should suddenly cease because of inclement weather, it's best to close up the hive and leave as soon as possible.

Photographs often show a beekeeper in the field, fully clad in suit, gloves, helmet, and veil. Being so protected is sometimes appropriate, but not always, especially when it's hot. Some beekeepers wear little or no protection, but that is not recommended either. The best advice is to wear what makes you feel comfortable and at ease around the hive. This can vary from occasion to occasion; experience will be the key.

Honey bee colonies are also sensitive to vibrations. Do not bang on the hive or jar the bees. They might overlook a single instance, but doing it again is a mistake. Other repetitive vibrations coming from lawn mowers or cars and trucks also can set off a colony's defensive behavior.

Bees have good and bad days. If there is trouble each time a colony is visited, step back

THE TAO OF SMOKING BEES

Some readers may be familiar with a martial art called tai chi chuan, or "ultimate fist," based on the Tao philosophy. A principle of this form is that one should strive to be like water, at once the softest of materials, but at the same time one that can erode rocks. Masters seek to be so soft that attackers' energy actually rebounds against them and they can be knocked down with little perceived action by the defender. Another way of putting this is that while less is more, much less is much more.

One Costa Rican scientist, Dr. William Ramirez, believes that beekeepers often oversmoke their colonies in his country and elsewhere in the American tropics. And when Africanized bees are treated this way, he says, all hell breaks loose. In the tropics, not only is an oversized smoke pot used in an attempt to keep these sometimes overdefensive insects under control, but there may also be one employee whose only job is to make and apply more and more smoke.

Use less, not more, smoke. This is Dr. Ramirez's rebuttal to those who would counsel such a course of action. A key part of this is to wait at least two minutes after administering the first puffs before beginning manipulations. This initial time period in fact may be the key to the Tao of smoking bees. It is possible that once that time threshold has passed, a switch has turned off the colony's defensive tendencies. Puffing and going in immediately, as is the habit of many beekeepers, may not allow enough time for the colony to get the message.

Much more study is required on this issue to develop better scientific knowledge about honey bee reaction to smoke. The defensive threshold may be quite different at certain times or with specific races of bees. The smoker remains the first line of defense, and the dictum to keep it well lighted and at the ready should always be heeded by the careful beekeeper.

and analyze what might be the matter. What is the weather like? Is the colony bothered by animal nuisances? Enlist another beekeeper to come, watch, and give pointers on technique. In general, treating a colony like a good friend brings dividends: the bees will often respond in like manner.

DRESS FOR SUCCESS WHEN MANAGING HIVES

- Wear light-colored, smooth fabrics
- Avoid red or black, which can trigger defensive responses
- Don't wear sweaters, flannel shirts, cotton athletic socks, and other fluffy fabrics. The insects' delicate hooked feet can be permanently caught in these materials
- Tie or tuck in long hair
- Tuck pant cuffs into high boots or socks
- Remove rings, watch, and other ornaments. Their movement might glitter in the sun, provoking defensive behavior
- Don't wear odors that might attract the insects, such as aftershave, hair spray, or deodorant
- Avoid other products that may also be offensive, including gasoline and petroleum products that give off fumes

The Beekeeper as Manager

Most books on beekeeping emphasize seasonal management for honey production. Seasons can have quite different characteristics, however, as a glance at the U.S. Department of Agriculture (USDA) Plant Hardiness Map opposite shows. Nectar production will vary as well.

Maximum honey production in the United States occurs in USDA Plant Hardiness Zones 2 to 6, where a relatively long, often harsh winter is followed by a prolonged growing season. This is due, in part, to increased day length during summer. In northerly climes, the bees must make large amounts of honey to survive the harsh winter, which accounts for a generally greater honey harvest. These weather conditions are less extreme in the subtropics and tropics, Zones 7 to 10, where seasonal rainfall instead of day length and temperature takes over as the major driving force in nature.

The apicultural calendar described beginning on page 109 reflects the events in the natural cycle that are correlated to honey bee population growth and managerial tasks by the beekeeper. It reinforces the bee manager's dictum that "all beekeeping is local."

The goal of the beekeeper producing honey in temperate regions is to help the colony develop a peak population of worker bees that coincides with maximum nectar secretion by plants. The commercial pollinator also needs a good population of foragers to ensure adequate coverage of plants in bloom. The best beekeeper monitors the colony but allows the bees to make most of the decisions about their own well-being, attempting to assist the bees only when they go off track.

One beekeeping axiom holds that if there is doubt about what you should do, the best

USDA HARDINESS ZONE MAP

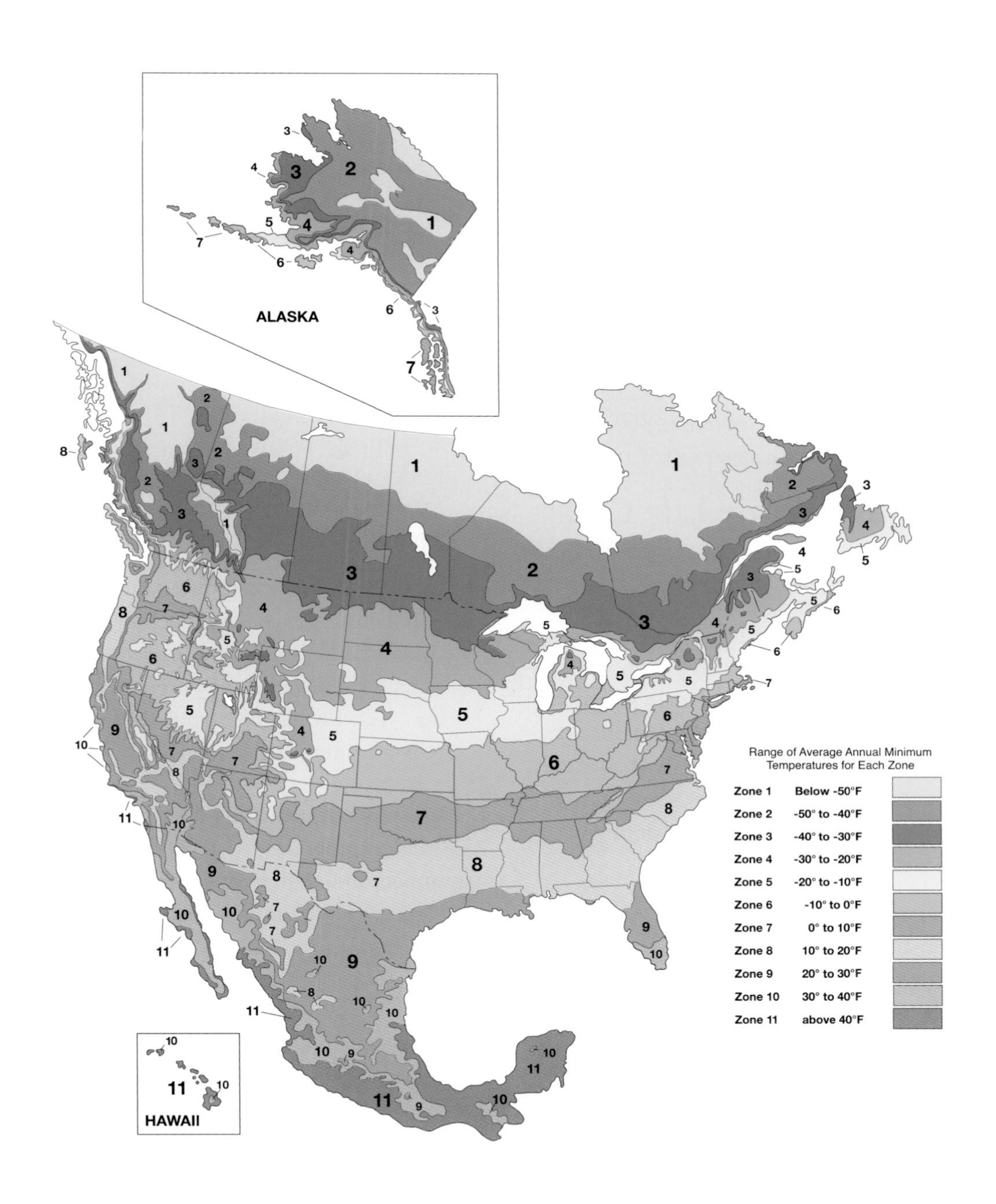

AVOID NIGHTTIME INSPECTIONS

Some of the worst encounters are not with flying bees, but with crawling bees at night. Instead of flying in the dark, they may crawl up from the ground or otherwise get under your clothes or veil.

If you must manipulate a colony at night, use a red light — bees are not attracted to it as they are to white light. Some beekeepers manipulate the bees in the dark using a red light if the daily temperature is extremely high. That allows them to work at a lower temperature, but they must still contend with crawling bees.

strategy is to do nothing. Waiting and observing is often better than hastily doing something that you might regret later. That said, there are times when it is imperative that the beekeeper step into the fray as manager of last resort.

A Colony's Yearly Life Cycle

Following is a description of the life cycle of a colony.

1. As the active season begins, the queen is stimulated by increased pollen and nectar resources to begin laying large numbers of eggs. This is a balancing act: the colony must continuously ensure the requisite number of workers are present to warm and care for the developing brood.
2. As population grows, the hive becomes constricted for space and drone eggs appear. The overcrowded colony eventually begins swarming preparations.
3. The colony constructs queen cells and rears queen replacements.
4. The old queen is readied for flight and the scouts begin to seek out new nesting sites.
5. Led by the old queen, the swarm issues on a warm day and establishes itself in a temporary bivouac before it decides to move to a new location.
6. A virgin queen is allowed to emerge in the parent colony and fly in order to mate so she can again develop a stable population.
7. Original and new colonies cycle through population increase and decline during summer and fall, depending on conditions. At year's end, they prepare to winter over with a stable population, ready to begin anew next year.

In temperate areas, all honey bee behavior is ultimately programmed to achieve a surplus in numbers and, therefore, a sufficient force to make honey. Population growth is influenced by conditions inside and outside the hive.

Population Cycle

A honey bee colony will fluctuate considerably in population size throughout the year. The average colony size is about 30,000 individuals, although it may be as high as 70,000 during strong nectar flows in the active season or as low as 10,000 in the middle of winter in temperate regions.

FIRST-SEASON POPULATION GROWTH FOR A COLONY INSTALLED IN APRIL IN A TEMPERATE ZONE

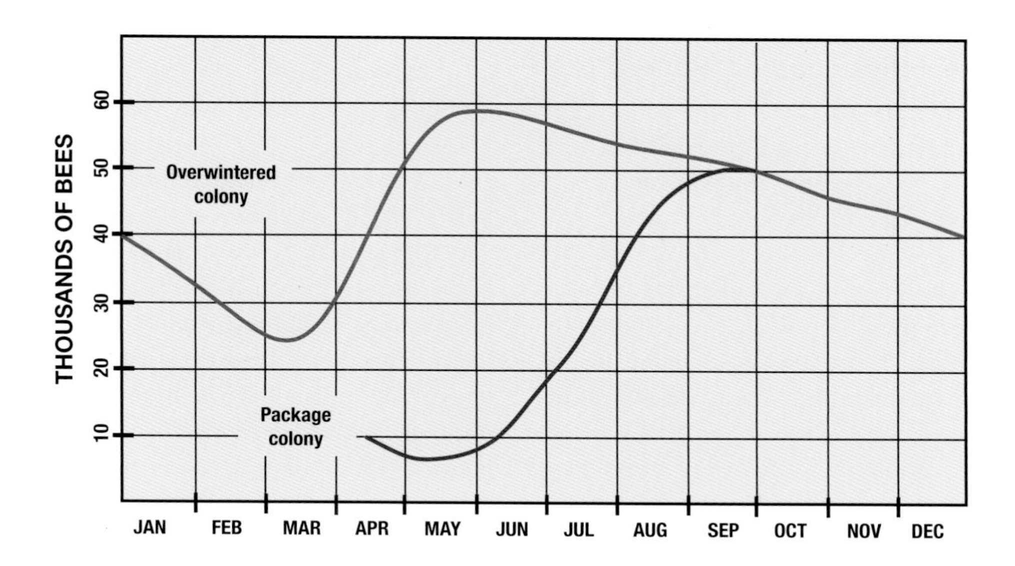

This graph reveals the difference in population growth between a colony installed as a package in April (in a temperate zone) and one that has successfully come through a winter. The package passed through a risky time of population decline just after it was installed. Only by September has it caught up with the overwintered colony, whose population is beginning to wane.

External Influences on Population

Outside conditions are often more apparent to the beekeeper than conditions inside the hive. Perhaps most obvious is ambient temperature, which can stimulate brood rearing and cause bees to gather water for evaporation if it's too hot or cluster together to produce heat during cold weather.

Day length influences both daily temperature and resource availability. Also called **photoperiod**, day length is a more reliable or conservative signal to living things than ambient temperature, which can vary widely from one day to the next. Nectar and pollen resources are also affected by temperature and photoperiod, as well as by many other factors such as relative humidity, soil type, and soil moisture retention.

By contrast, in more tropical areas ambient temperature and photoperiod are less influential, and moisture availability (rainfall) plays a greater role. Honey bees in these areas are not compelled to produce large amounts of honey to survive harsh winters, and plants also are not adapted to produce the huge nectar crops found in more northerly climates. Rather, the goal of population growth is reproduction, and swarming and absconding take on greater survival roles than honey production.

Internal Influences on Population

Internal factors affecting a bee colony are many. Absence of a laying queen indicates a population is in decline. The population balance among queen, worker, and drone will determine what duties are carried out by which contingent of worker bees and when. The varroa mite affects the bee colony's population growth in several ways, including direct parasitism and transmission or activation of viruses.

With population growth, the efficiency per bee increases, and food storage becomes a major activity in temperate regions. Workers collect protein in the form of pollen and consume it to produce jelly to feed the huge number of larvae hatching from eggs. Nectar is collected and processed into honey at a rapid rate. As the colony increases in size, the bees begin to sense that overpopulation is imminent, and they make preparations to swarm.

After swarming, the population may slowly recover as the honey flow continues. At the end of this yearly cycle, the bees achieve a steady state of population for maintenance, and the cycle is ready to repeat once again.

A similar cycle is seen in more southerly latitudes, but it can be based on different circumstances. Honey production is not as important a survival tool as the ability to move away from pests and migrate in search of water. Although individual colonies may produce less honey in the tropics (an oft-heard complaint about Africanized honey bees), the beekeeper can keep a far greater number of colonies than would be possible with European bees in the same environment.

SPLITTING HIVES

There are many ways to split a colony of honey bees. It's relatively easy to divide the brood that is present between two colonies. The difficult part is ensuring that each has an adequate population of adults. This is usually accomplished by putting more adults into the half of the division that is moved to a new location, because many will fly back to the old nest location. As a corollary, leaving the weaker half of the split colony at the old location will also result in a larger population. Older foraging bees from the relocated part will fly back to the original location.

Anywhere from four to six frames of sealed brood are usually needed to make new divides. Brood close to emerging can be placed in units where adult populations are low, because fewer nurse bees are needed, unless the nights are very cold and there is risk that, unprotected, it will chill and die. Emergence of this brood shortens the time before the new divide can again reach full strength. In cases in which bees are likely to swarm anyway, making divisions becomes a good way to use the honey bee's reproductive instinct to the beekeeper's advantage.

Success in making divides depends on the beekeeper's judgment and experience. Even the most experienced operator makes mistakes in some years. The bees themselves are not immune to making the same error, so the beekeeper must make a more rational decision at times.

The Apicultural Calendar

Beekeeper and bee activities follow what is called the apicultural calendar, usually divided into "active" and "inactive" times for the colony. In temperate lands, the four seasons are well defined, but in the subtropics, this distinction becomes less clear. The Mediterranean climate of California, for example, has two seasons, a dry summer and a wet winter. As one approaches the equator, it is the opposite, with a wet summer and a dry winter. There the nectar flows generally occur in winter, as opposed to spring in more northerly regions.

The calendar sequence below describes the temperate climate found in the midwestern and northeastern U.S. For specific beekeeping situations in the subtropics and tropics — or anywhere you are — consult your local beekeepers' association or folks with many years' experience in the area. Colony population growth correlates with the kinds of plants that might be available, and this in turn affects beekeeper activity. In temperate regions, the bee population rapidly expands as pollen-producing plants bloom. At this time, beekeepers check for brood and stores, feeding if necessary. Next come the prolific nectar-producing plants that make up the **honey** (or **nectar**) **flow**, signaling the beekeeper to add supers, control swarming, and finally remove the honey crop. This transitions into a leveling off of the population in summer, and even a decline if the colony swarms.

In many areas, another time of increased activity might follow, correlating with Indian summer and lower temperatures. This can result in a late summer or early fall honey flow, which can be substantial in some regions but is not usually as reliable as the one in spring. The colony and the beekeeper then prepare for the coming inactive season. At this time, varroa control and queen egg laying are critical to produce the vital winter bees the colony needs to survive this harsh season.

Varroa Cycle

Now that the varroa mite is part of every colony, we also must take its cycle into consideration as part of the apicultural calendar. In the early part of the active season, a honey bee population can outstrip varroa as the colony rapidly builds in number. When the colony reaches a peak, however, and begins to decelerate, the varroa population will catch up and often continues to rise as the bee population levels out and then declines in advance of the inactive season. In essence, the brood becomes a mite factory. It is at this time that varroa is most dangerous and monitoring its population is most critical.

Spring Management

In temperate lands, bees and people look forward to spring. It is when the beekeeper must "spring" into action, marking the end of the winter doldrums. In spite of the optimism, however, it is important not to rush the season. It can be easy to do this, especially in the North where so-called cabin fever might have set in.

March in the southern United States and April in the northern regions — the beginning of spring in USDA Hardiness Zones 7 and 6, respectively — are critical seasons. Things can change quickly. It's difficult to know when to inspect the bees at this time. You may have a warm and sunny spell but cool, rainy, unpleasant days are yet to come. A good sign to look for in the North is the dandelion bloom; in the South it is the blooming of the swamp maples and willows. Both are producing pollen, which the bees need to begin or sustain brood rearing. As the season progresses, nectar plants begin to predominate.

Estimating the amount of brood in a colony is a key management skill.

RECOGNIZING A STRONG COLONY

"Strong," "marginal," and "weak" are relative terms used by beekeepers to assess colony condition based on population size. It is common wisdom and common sense that colonies should be kept strong for maximum productivity. Consider the words of L. L. Langstroth, the man who revolutionized beekeeping and invented the hive most commonly used today, writing in 1859:

> *The essence of all profitable beekeeping is contained in Oettl's Golden Rule: Keep your stocks strong. If you cannot succeed in doing this, the more money you invest in bees, the heavier will be your losses; while, if your stocks are strong, you will show that you are a beemaster, as well as a beekeeper, and may safely calculate on generous returns from your industrious subjects.*

Colony strength generally reflects local climatic conditions. The USDA Plant Hardiness Zone map (see page 105) can be roughly divided into two parts with specific conditions affecting honey bee colonies:

The **temperate region** generally encompasses Zones 1 through 6, with an average minimum temperature range from –50°F to –10°F (–45.6 to –23.3°C). Longer summer days allow more foraging time for honey bees than in areas farther south. The highest honey production tends to occur in this region. Winters are longer and colder, however, with greater risk of colonies perishing then.

The **subtropical region** generally encompasses Zones 7 through 11, where the average minimum temperature ranges from –5°F to above 40°F (–20.6 to 4.4°C). Summer days are shorter and honey production is thus not as high as in the temperate region. In contrast, winters are milder. Honey bees in this region are less likely to perish from starvation but may be more vulnerable to pests and predators, especially varroa mites.

Spring: First Inspection

The initial spring inspection of overwintered established colonies should be brief. As with managing package bees, keep certain objectives in mind for each visit. You'll want to determine the status of the queen, estimate the population numbers of worker bees, evaluate the amount of brood and food, and look for evidence of disease or pests. Assessing these takes time and experience. Food is most critical; when in doubt, feed.

Amount of brood. Determining the amount and condition of brood (along with the size of the worker population at the beginning of the active season) is a key skill for beekeepers, who often refer to numbers of frames of bees and brood. If more precision is required, you can make a template from wires in square inches to estimate the amount of brood present.

It is also important to notice the shifting ratio of eggs, larvae, and pupae. Early in the season, young uncapped brood predominates but lessens progressively over time as the amount of capped brood increases.

Condition of brood. The condition of the brood is also important. Caps over brood should be uniform in color, and the best overall pattern is a solid mass of brood interspersed with only a few empty cells. **Capped brood** also covers and protects mites; the more of this brood there is, the more the potential varroa population in the future.

ZONES 1 TO 6

The following conditions show a "strong" colony in the temperate region in early spring:

1. Bees occupying 12 to 15 frames in the brood chamber.
2. Brood in 6 to 10 frames in the brood nest.
3. At least 20 pounds (9 kg) of honey: four full frames.
4. At least 2 to 3 frames with pollen.

These guidelines are for a hive with a brood chamber made up of two standard hive bodies and an appropriate number of honey supers. This is the most common setup in the temperate region.

ZONES 7 TO 11

Southern bee managers generally use only one brood chamber and a shallow super or food chamber as their basic unit. The smaller winter population demands less space. The following is, therefore, optimal for a "strong" colony in the subtropical region in early spring:

1. Bees occupying 8 to 10 frames in the brood chamber.
2. Brood in 4 to 5 frames in the brood nest.
3. At least 15 pounds (6.8 kg) of honey: three full frames.
4. At least 1 to 2 frames with pollen.

Varroa population. Estimate the varroa mite population during every inspection. This is not easy: it often fluctuates greatly depending on conditions. Sometimes the mites outpace their host; other times it is the reverse. Whether the mites are mostly found in brood or on adults, for example, can make a difference. At the beginning of an infestation, before viral epidemics from the mites have become critical, treatments can be worthwhile. Waiting too late to treat often dooms a honey bee colony.

If all seems to be going well (again, not always easy to figure out), you can schedule the next inspection with some optimism. That should occur within 30 days.

Spring: Subsequent Inspections

Subsequent inspections — which should take place, on average, at least once every two weeks — should more or less mirror the initial one. Again, the paramount goal is to estimate the population of workers, the amount and condition of the brood, and the quantity of food. If the varroa population is low but the colony still does not seem to be progressing, is under normal strength, or appears to lack vitality, it's time to consider intervening.

Before acting, however, it is important to figure out what condition is most important to correct and what resources are available to do so. If population growth appears to be a

BEEKEEPER'S STORY

MY INTEREST IN BEEKEEPING was generated by a long family history of beekeepers. My great-grandfather, Fletcher Hall, began beekeeping in Mason Valley, Nevada, in the 1920s. In 1926, my grandfather joined the business. My father, Harold Hall, was 6 months old at the time. The business grew and later became known as Ralph Hall and Sons Inc., or locally as Hall's Honey. My grandfather, and later my father, were appointed by Nevada governors to the Nevada Department of Agriculture Board of Directors as representatives of the beekeeping industry and together served for more than 40 years.

In 1976, Ralph and his sons, Harold and Walter, sold one of the largest beekeeping businesses in Nevada. Even though I grew up in the business, I was young, involved in school and 4-H activities, and didn't pay attention to it. I was employed during the summer months as an extractor operator. In 2008, after hearing of the decline of honey bees, I decided to start a few colonies and to continue the family tradition. I discovered that the Nevada Board of Agriculture has no beekeeping representative nor any apiculturist on staff.

The first time I experienced inspecting my own beehive, it seemed a little intimidating. It didn't take long to realize how calm the bees were, and then it was all excitement to see the process and what they were doing within the hive.

Debbie Gilmore, Nevada

problem, additional bees might be considered. With only one colony, the options are limited, which is the reason this book recommends novice beekeepers begin with two colonies. If you have only one colony, seek out a fellow beekeeper for help, or turn to a commercial firm producing package bees, queens, or nucs for reinforcement bees.

If the colony appears to be weak despite low varroa numbers, there are some possible actions you can take. The following are listed in order of preference.

1. **Add a frame or more of capped brood.** Capped brood requires minimal care and will not be a drain on the colony's nutritional reserves. It will also result in new workers more quickly than younger brood. Inspect the brood for diseases and pests before adding it.
2. **Combine with another colony.** Some colonies are not vigorous enough to develop into productive units. These chronically less-than-optimal units (sometimes called "dinks") are a headache to the beekeeper, magnets for diseases and pests, and prime targets for robbing. Combining two weak colonies often increases their potential to confront these problems. Many beekeepers are loath to take such an action, but it's often better to cut one's losses and save the bees and their resources rather than let a colony dwindle.
3. **Add a small package with or without queen.** Although this provides a population ready to go to work, there are several potential disadvantages. The additional bees will be seen as strangers, and some fighting among workers might ensue. Any queen in a colony is at extreme risk of being killed by bees that are added, because her odor does not match the one they are used to. If, however, the unit is deemed queenless, adding bees and indirectly introducing a new queen might be just what is needed.
4. **Replace the queen.** If the queen's brood production is suspect, replacing her becomes an option. Too many open or empty cells in a brood frame otherwise filled with capped brood are one sign. This is called "shotgun" brood and can be an indication of disease or a failing queen. If there are too many drones or a surplus of drone brood, the queen may have run out of semen. A young queen can inject a new sense of purpose into a colony. Adding a frame of capped brood at the same time will further jump-start population growth. Although there are few hard-and-fast rules in managing bees, one has stood the test of time: when in doubt, requeen.

As inspections continue and the population builds, the beekeeper turns to other concerns. These include adding room to a colony so that population growth is not impeded and in the hope of preventing the ever-present possibility of the bees preparing to swarm.

Spring: Supering

When bees are found to cover most frames in a colony, the best advice is to add more supers of drawn or foundation combs. Depending on location and beekeeper preference, this might be in the form of a deep (full-size) or shallow super.

In the North, where two full-size supers usually make up the brood chamber, beekeepers simply reverse their order, removing the usually empty bottom box and placing it on top of the colony. During cold weather, the cluster of bees slowly moves upward in a colony covering the food supply and replacing it with brood when possible. At the end of the season, most of the bees and brood are in the top. Reversing the supers puts the majority of bees and brood in a lower position, leaving room at the top for expansion. Timing the reversal is extremely important. If it's done too soon, the colony becomes stressed; if done too late, the colony becomes too crowded, which promotes swarming. Once the two brood chambers are populated, honey supers are added as needed.

In southern regions, where the overwintering unit is a deep super with a shallow or "food chamber," the winter cluster is not organized for long periods of time. The bees have more flexibility and the practice of reversing appears to have limited value. As room is needed, more shallow supers are placed on top of a colony.

There are two ways to add additional supers to a hive. They can either be stacked on top (**top supering**) or inserted underneath the top super (**bottom supering**). There are continuing discussions among beekeepers about these with little agreement. What is key is providing room judiciously; too soon and the bees ignore the supers; too late and swarming is the result.

Spring: Controlling Swarming

Swarming must be a marvelous time for honey bees as they seem to revel in their fervor to reproduce. Many a human being has been caught up in the excitement, too. Poems and songs have been written to commemorate this behavior, and most beekeepers will relate with a good deal of passion their first experience with swarming bees.

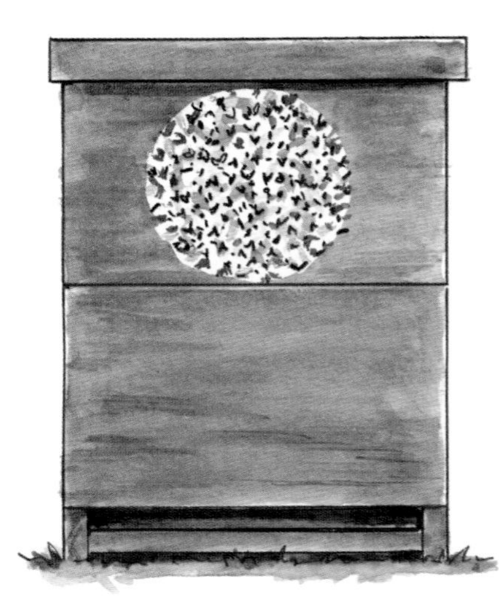

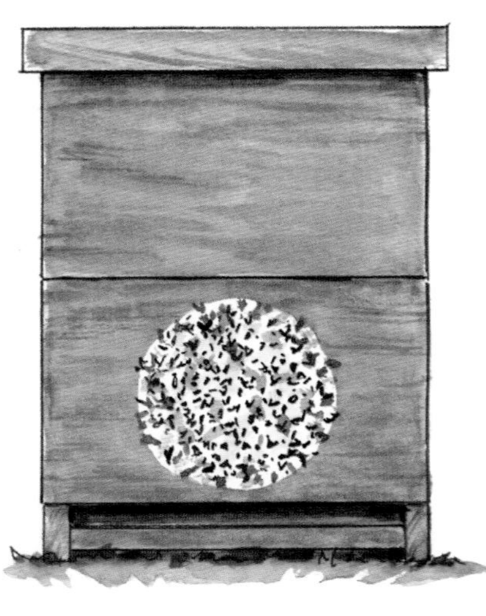

Reversing hive bodies in spring usually provides the honey bee colony with a new lease on life. At left, the colony has concluded the winter with the cluster located in the top body. There is literally no place for it to go from there. When the two bodies are reversed, as at right, the cluster is now in the bottom of the hive, providing heat as well as space for upward expansion.

There is, however, a big downside to swarming, for bees and beekeeper. The colony's excess honey has flown away in the collective crop of the swarm of departing bees. This could be a disaster for the swarm if it cannot soon find suitable forage to add to its food supply. In addition, the bees will lose much productive time, waiting for their queen's progeny to emerge.

At the same time, the parent colony can suffer. There may not be time to regenerate the honey lost. Some colonies may not be satisfied with a single swarm. Virgin queens could continue to emerge, and cast what are called "after swarms," further weakening the unit.

Because swarming is the bane of any beekeeper attempting to produce honey, it shouldn't be a surprise that the topic has merited much attention over the years. The urge to reproduce is so strong that sometimes it seems bees will swarm no matter the circumstances. Nevertheless, beekeepers continue to try to keep this fundamental urge under control.

The following conditions and observations may be a tip-off that swarming is a strong possibility:

- Winter was mild
- Spring is early
- Plentiful food reserves are on hand
- Queen cells are observed
- The queen is more than 1 year old
- Drone rearing is early

The amount of sun, shade, and ventilation a colony is exposed to should also be taken into consideration. The best time to engage in swarm control is before the colony "gets it in its head" to leave. The best strategy to employ is providing room for expansion to reduce crowding in the nest; the earlier the better. The following techniques are often recommended.

LAST-MINUTE SWARM PREVENTION

Usually when the colony begins rearing queen cells, it means the swarming instinct has been triggered. Two more drastic procedures are then sometimes recommended, but they are risky and often unsuccessful:

1. **Destroying queen cells so they will not develop.** Bees will usually not leave behind a potentially queenless colony.
2. **Removing or caging the queen**, which results in fewer workers being produced and less congestion.

The following techniques involve preempting the swarming activity of colonies by reducing the population pressure in a colony:

1. **Splitting the hive** that is about to swarm and making up a new one. The resulting two colonies will not be in a swarming mode anytime soon, if ever, that particular season.
2. **Removing frames of brood and/or house bees and adding them to weaker colonies. Capped brood is** generally the better option. Honey bees will readily accept brood, whereas if adults are added to existing colonies, conflict will occur unless the bees are newly emerged and introduced slowly over time.

Manipulate frames to provide egg-laying room. Moving center frames full of brood elsewhere and replacing them with empties is a time-honored technique. One variant, which involves separating the queen and foraging bees from the brood and nurse bees, is called "Demareeing," named after its inventor, George Demaree.

Add supers to provide more room for storing and processing honey. Keep doing this over time. Again, it is better to err on the side of adding room earlier than later.

Insert foundation. New foundation provides an outlet for the wax production by young bees.

Move colonies to equalize population strength. This is another technique whereby strong and weak colony locations are switched. The field bees from the stronger colony beef up the population of the weak one.

Genetic Influences

One reason the prevalence of swarming is high is because, over centuries, beekeepers have encouraged the activity in honey bees. Before the movable-frame hive came on the scene, it was the only way beekeepers could increase their colony numbers. Some varieties of bees, therefore, are genetically programmed to be "swarmy." One potentially long-range strategy would be to select bees who do not exhibit the behavior. Some beekeepers have been successful in this by simply requeening any colony that swarms with queens raised from bees who are known to be nonswarmers. Any captured swarm also gets the same treatment.

Summer Management

The peak bloom for many nectar plants is early summer in the northern United States. When this happens, it's time to see whether all the bees' and your hard work getting populations up to their optimal level will pay off. Don't be surprised, however, if it doesn't work out the way you planned it. Conditions vary greatly from year to year. It takes several years' experience in a location to get a sense of how consistent it will be. The litany of potential problems is long: it might be too hot, too cold, too dry, too windy, too wet, and so on. Bees and the vegetation they depend on can be affected greatly by shifting environmental conditions, even on a day-to-day basis.

Visits to the apiary in summer are made mainly to look at storage issues. Plan to visit every two weeks. It may be time to add and perhaps take off supers as previous ones become filled and capped. Some beekeepers begin to extract honey early; others wait until the season is over. As part of this, continue to monitor

the adult population. A good way to estimate what's going on is by using a hive scale to monitor weight changes.

In the South, the major nectar flows may be over, having peaked in late spring. Summer in this region can be extremely hot with no appreciable nectar. In some locations, colonies can even be hungry, especially if the beekeeper has been too anxious to take off a honey crop. Afternoon thundershowers can wash away a nectar flow quickly, and the bees might become prickly without warning. Always be ready (with a smoker and your protective clothing) at this time for unexpected defensive behavior. Robbing can also begin without notice, increasing defensiveness.

Other concerns, aside from the ever-present threat of swarming, include the possibility of one queen taking the place of another and the condition of the frames and combs. Queens are hard to find when populations are large, and looking for one at this time of year is not recommended if the population appears adequate. If the number of bees becomes reduced, however, it is reasonable to take a closer look at the brood and queen. If the queen was marked and one appears that isn't, this means a **queen supersedure** has occurred (see page 118).

Damaged frames and combs should be identified for replacement; then they can be slowly worked from the center of the nest to the sides and eventually removed. Total comb renovation represents a revolution in thinking; older literature often reflected beekeepers' beliefs that even very old comb was serviceable. This is no longer the case. Use of varroa treatment chemicals inside the beehive can contaminate the wax, which might become toxic. Many beekeepers are now going to three-year rotation for all combs. Some have abandoned beeswax foundation entirely in favor of plastic, which does not become contaminated and can easily be renovated by the bees after the old wax is scraped off. (See page 164 for more on comb renovation.)

There are many possible ways to monitor weight change in a colony. Here, the beekeeper uses a modified hand truck attached to a sensing device. This technology has become more common in recent years for other applications, such as studying climate change's effects on bee forage.

SUPERSEDURE

Queen supersedure occurs when the colony determines the queen is failing in some way. The bees construct a specialized queen cup on the face of the comb and the queen is encouraged to lay an egg in it. After the new queen emerges and mates, the old one is eliminated. Queen supersedure creates a break in the brood cycle, thus lowering potential population growth.

This supersedure cell cup on the face of the comb is ready to produce a replacement queen.

Late-Season Management

The late season really begins in late summer: July and August, to be exact. At this time, the beekeeper begins to anticipate and prepare for the coming winter. In the North, USDA Plant Hardiness Zone 6, the target date is August 1. A month or so later is appropriate for beekeepers in Zone 7. This is a critical season. There may be a late honey flow in various regions, complicating things in unexpected ways, such as restricting the queen's egg laying.

The objective of management at this time of year is to ensure that a viable population of honey bees goes into winter with a good chance of surviving. Young bees are important, but even more significant are good healthy populations of "winter bees." These overwintering insects are adapted to storing nutrients for a long period of time. Summer bees cannot do this as they lack well-developed fat bodies.

The origin of this vital population of winter bees is the queen. The beekeeper must, therefore, take pains to ensure she is up to the job. There is a natural slowdown of brood rearing at this time, so a failing queen may not be detected by the beekeeper. On the other hand, honey bees are good at preparing, and one may see signs of supersedure cups being constructed, meaning a new queen is on the way.

If there is any doubt about the queen's condition, serious thought should be given to requeening. Some beekeepers requeen in late summer or early fall on a regular annual basis. New queens lay at a higher rate than older ones. Late summer requeening allows for multiple chances for queen acceptance. The resultant population is larger. In addition, a first-year queen is much less apt to swarm the following spring.

A substandard population in the late season can also hinder a colony's preparations for winter. Combining weak colonies into a stronger unit that has a better chance to survive the coming harsh conditions is a time-honored approach. The axiom "Take losses in the fall" applies. Many beekeepers are loath to reduce their colony numbers, however, and often wait too long. If colonies are combined into stronger units and survive, there is always the opportunity to split them in spring.

The late season is the most important one for varroa management. Parasitized honey bees are not good candidates for winter survival. There is usually a large mite population in the late season, fueled by all the brood the colony produced since beginning the active season. Many mites are protected in the brood cells and are not susceptible to chemical exposure. A break in the brood cycle at this time may be desirable to let brood emerge. In the process, female mites will have a reduced brood population available to be parasitized, and the mites are vulnerable to the presence of any chemical control.

Brood nest management for mite control is something that is not emphasized in books written before varroa was introduced. Requeening can also produce the brood cycle interruption needed to control mites. So can dividing strong colonies, treating and requeening the splits, and then letting them overwinter. This technique has been employed by a number of beekeepers with surprisingly good results in moderate climates. In harsher climates, the same management can be used and the resulting smaller units can then be overwintered in a shed, garage, or cellar as nuclei. (See chapter 10 for more on treating varroa.)

Winter Management

Winter management begins in late fall. The honey bee cluster forms at 57°F (14°C). Once that occurs, little can be done to manipulate hives with any degree of success. It is during this period that the beekeeper turns to repairing equipment and getting the extraction area shipshape for the coming active season.

Some things may deserve consideration as the cold weather descends. It could still be possible to ensure colonies are protected from rodents and the elements. In areas where mice are present, install mouse guards to prevent them from entering hives when the bees are clustering above the bottom board. Bales of hay or the odd piece of plywood might be employed as windbreaks to protect hives from fierce winds.

There is considerable debate about wrapping colonies with tar paper, straw, or other materials, as is done in some regions. It is important to understand that honey bees do not attempt to warm the entire space of the hive, but only the discrete cluster. The hive should never be closed entirely. The cluster produces a great deal of warm, humid air that must be exhausted. Providing an upper entrance during winter facilitates this. If this air is not removed, moisture may condense on the inner surfaces of the hive, which is highly harmful to bees. Sometimes the outer cover is slightly propped up by small wedges to increase ventilation.

Protecting colonies with a top insulation board is also possible. There is no airspace between it and the inner cover, which has been turned upside down. Some beekeepers simply put an empty super filled with a fiberglass-type material on top of a hive. This material must be protected by a screen or the bees will attempt to remove it. The use of supers filled with hay, straw, or leaves is not recommended because these materials can trap a lot of moisture.

Since beekeeping began, managers in temperate areas have experimented with wintering indoors. In certain regions, especially the plains of Nebraska and prairies of Canada, this is accepted practice. Colonies are made into nuclei and then stored indoors in the dark at a constant temperature of around 40°F (4°C). Gases produced are exhausted, and the colonies are then stored in an artificial quasi-hibernation state until spring.

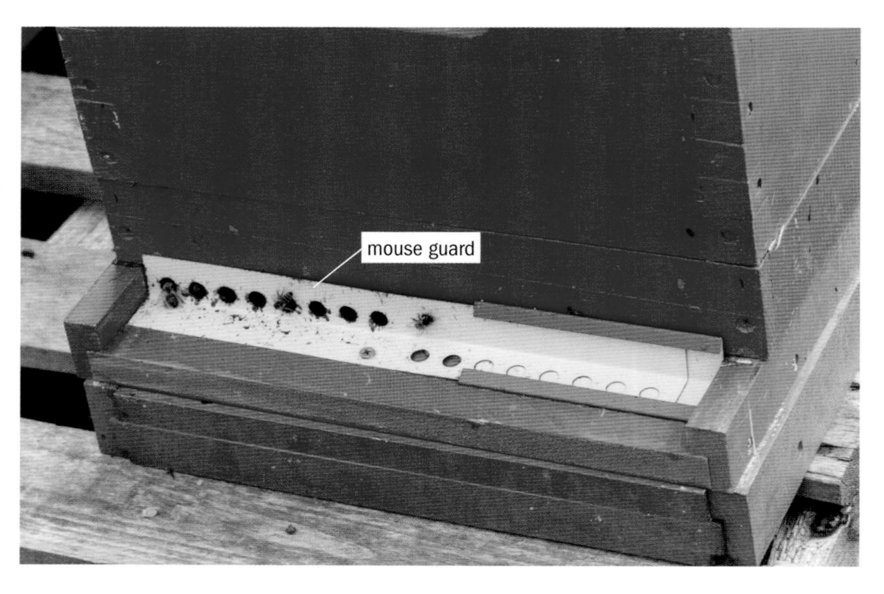

Winter preparations include installing mouse guards (you can use ½-inch hardware cloth), insulating (adding a board or wrapping colonies with paper or straw), and ensuring there is sufficient ventilation to let out warm, humid air.

Requeening

The ability to change a single individual, the queen, and thus influence the behavior and direction of a honey bee colony, is a great boon to the beekeeper. One of the universal answers to problems observed in a bee colony is, therefore, simply to requeen. Most often this solves not one but a whole host of problems.

All queens are not equal, however. A queen's egg-laying potential varies based on the conditions under which she was reared; her age; her lineage; her history; and perhaps most significant, the kind and number of sperm she carries.

Beekeepers can't judge much about a queen beyond knowing her age until they observe her resulting colony. One of the debates in beekeeping circles involves whether beekeepers should purchase queens or let the bees rear their own when making splits or replacing the current queen. There is no clear answer, although it makes sense that a commercial producer would raise a queen in an environment more controlled than is possible in a colony left to its own devices. The question, then, is whether producers are held to any standard. It makes sense that beekeepers would be better served generally by asking specific questions about rearing conditions, but many don't know what to ask, and there are few definitive answers if they do.

The Queen's Lineage

Perhaps the most important aspect of queens — lineage — is the most difficult to gauge. This determines to a great extent the eventual population she will produce. The makeup is complicated by many variables, including the fact that any queen may be mated with 5 to 40 drones. A queen's progeny, therefore, is a complex of subfamilies, the genes coming from a number of fathers. The fathers' influence is often dominant. As a consequence, a queen's offspring is assymetrical, made up of **half sisters** (with unrelated fathers), full sisters (whose fathers are brothers), and **super sisters** (sharing a common father). This diversity is the great strength of a colony. It ensures better resistance to pests and diseases, and colonies with greater diversity produce more comb, establish nest sites earlier, forage more efficiently, and overwinter more successfully.

Over many years, increased genetic diversity has enabled honey bee populations to take up residence in a variety of geographic settings, resulting in a number of discrete populations, or ecotypes. These Old World populations, however, were altered over time with their introduction to the Americas.

It is doubtful that any pure populations or ecotypes exist in the United States. The queen producer may mention that Italians are available, or Carniolans, or even a fairly recent type, Russians. Others may advertise Buckfast, Minnesota Hygienic, New World Carniolans, or USDA Mite Resistant. The latter four categories are not ecotypes so much as a collection of behavioral and genetic characteristics selected by certain breeders. Again, the beginner should at least ask the lineage of any queen purchased and how it was determined. The answer can be filed away in the experiential vault and resurrected when needed.

Both rearing conditions and lineage factors mentioned above are reflected in price. Many believe that beekeepers have generally given up quality concerns in favor of low-priced queens. It seems clear that as the problems in keeping colonies increase, and the more important a queen becomes as part of the eventual solution, the more expensive quality individuals will become in the future.

Rearing Quality Queens

Practically every time the discussion gets around to rearing queens, an argument erupts. There is little agreement about whether beekeepers should let the bees raise their own. The insects, after all, have reared replacement queens for millennia with obvious success. Why should beekeepers put out hard-earned cash for something they can get at no charge? Many beekeepers also complain that queens purchased from commercial outfits are quickly superseded.

Queen rearing on any scale is not easy. The hard work, monetary investment, and patience required are astounding. Much of the queen-producing season also takes place during marginal weather conditions, increasing the risk that many individuals will be lost in the process.

After some reflection, it boils down to quality. A quality queen is produced under rigorously controlled conditions, maximizing the end product's value. Obviously, some entrepreneurs do a better job than others. Judging quality is difficult as there are no published standards. It is believed, however, that in general the larger the developing larva, the better the product.

Rearing the best-nourished queen larvae is the goal of every producer. In the pursuit of quality, many cells with marginal individuals are discarded during the production process.

Because there are so many variables inherent in queen production, beekeepers often successfully let the bees produce their own queens.

It should be realized, however, that if selection is nonexistent and/or the conditions used to produce queens are marginal, the

MARKING A QUEEN

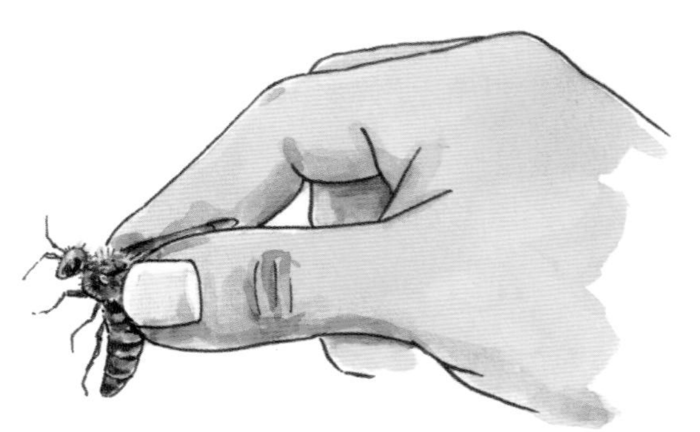

1. Carefully pick up the queen by the wings and thorax.

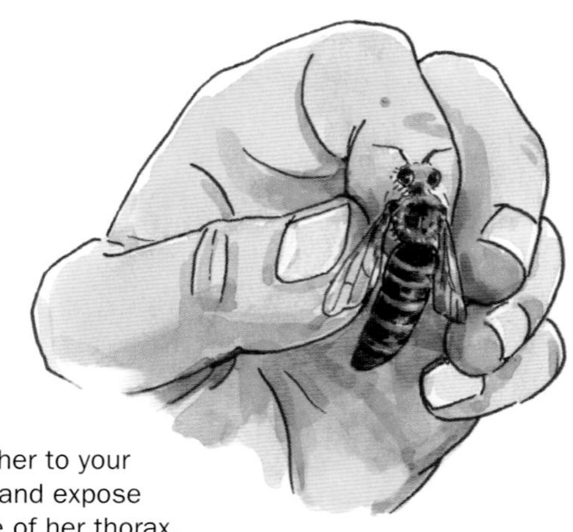

2. Transfer her to your other hand and expose the top side of her thorax.

possibility of ending up with a quality product is reduced. As with wine production, quality queens don't come easily. It is no wonder that many queen-rearing operations have had to mature over several human generations to ensure continuous production of the honey bee colony's most important individual.

Marking the Queen

A standard request from the beginning beekeeper to the supplier should be to ship a marked queen. If the beekeeper can't tell the original queen in a colony, it's impossible to determine if she's been replaced by the bees. Marking provides this advantage. Another benefit is that the queen is much more visible during inspection.

Should a queen be found who is not marked, she should be as soon as possible. This appears to be a daunting task at the outset, but after a little experience, can be quickly accomplished with little risk. She can either be immobilized on the comb with a special tool purchased at supply houses or caught by hand and immobilized. The latter is most common.

Pick her up by the wings and thorax, never the abdomen. Use one hand to pluck her off the comb and then transfer her to the other hand, being extra-careful when grabbing her legs. Apply the paint to the top of the thorax or bases of the wings. Do not get any on the abdomen or head! Let the paint dry for a minute or two before returning the queen to the nest.

Lacquer sold in hobby shops for painting models, typing correction fluid, and fingernail polish are all acceptable. Make sure it dries quickly, an essential attribute of any marking solution.

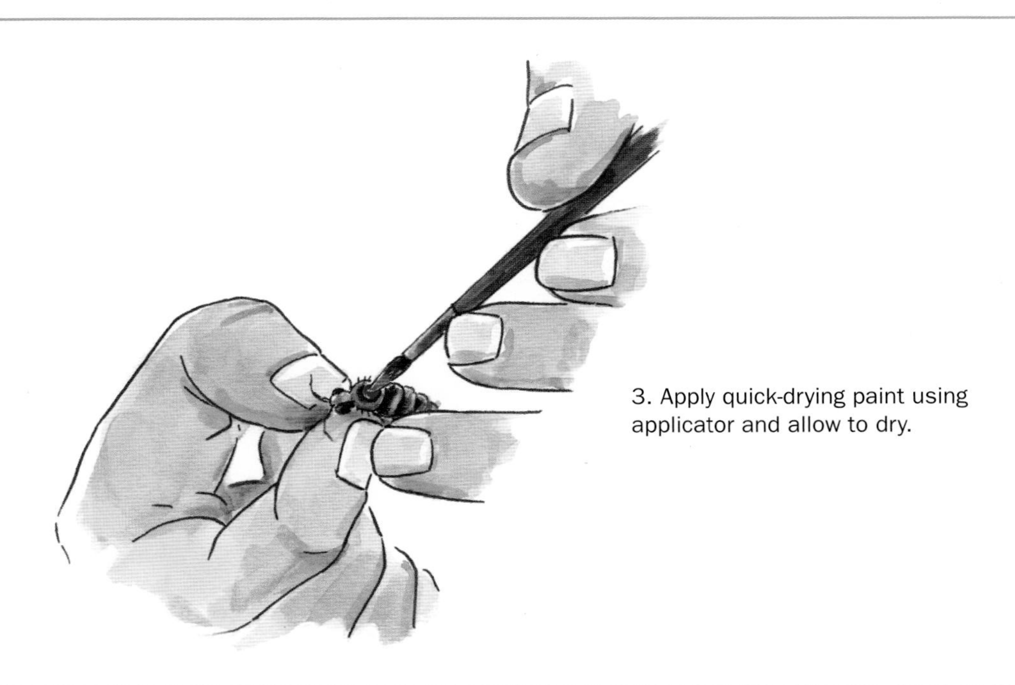

3. Apply quick-drying paint using applicator and allow to dry.

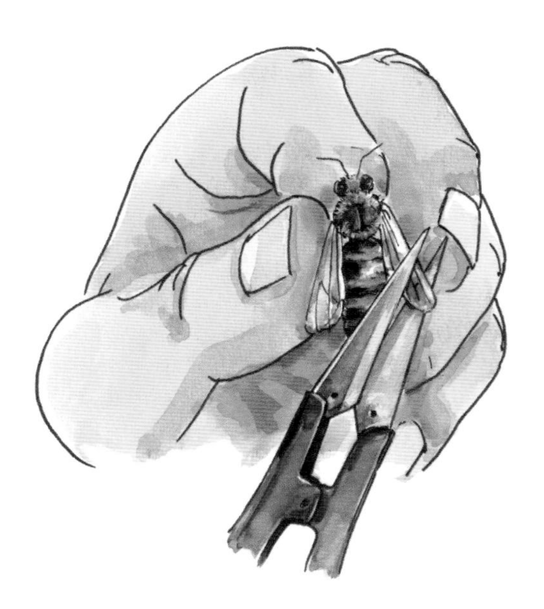

Do not clip more than one third of the wing.

Another possibility is to clip the queen's wing. It doesn't make her as noticeable as marking, but does keep her from flying off with a swarm. Once the bees notice the queen is not with them, they will usually return to the hive. The queen can easily become lost in the process, however. Clipping is usually done with a pair of sharp scissors. Again, she must be immobilized. Be sure not to clip more than one third of the wing as there are tiny veins that run through it.

Both marking and clipping can be done at the hive, but this is risky. An alternative procedure is to cage the queen and then move to a small room in a house like a bathroom or the cab of a truck with a bright window. If the queen flies off while the beekeeper attempts to mark or clip her wing, she will be attracted to the light and much easier to recapture off the glass.

The best training exercises for learning to handle queens consist first of gaining

The blue dot on this queen indicates that she was marked in a year ending with 0 or 5.

COLOR CODING THE QUEEN

There is an international color-marking protocol that has been published, although not everyone follows this convention. The last digit of the year determines the color.

Year Ends with	Color
0 or 5	Blue
1 or 6	White
2 or 7	Yellow
3 or 8	Red
4 or 9	Green

experience by picking up drones, then moving to workers. Drones cannot sting and are robust; they can take a lot more punishment than queens or workers. Workers are smaller than queens and are able to punish anyone handling them improperly.

Managing Nutrition

Honey bees are vegetarians and consume only plant juice (nectar) for their energy needs and pollen for development. Most beekeepers are aware that colonies cannot survive without honey or nectar stores. On the protein side, it is known that honey bees need 10 essential amino acids for growth and development. Pollen provides these in varying amounts. Certain pollens from fruit trees, for example, are much more nutritious than that of other species such as pines. This is reflected in the fact that bees are able to better collect "sticky" pollen from the former, whereas the latter is adapted mostly for wind dispersal.

Pollen also contains enzymes important to bee health and development, sterols for hormonal production, minerals, and vitamins. Pollen deficiency in workers has been found to adversely affect the brood food (hypopharyngeal) gland and result in reduced longevity. In queen bees, lack of protein and fat-body development results in early supersedure and failure of virgins to adequately mate. Its lack may also be responsible for a reduction in the number of drones available for mating and/or a decrease in their potential sperm production.

In spite of a rather long research history on honey bee nutrition, it is surprising how little is known about the subject. Effects of nutritional deficiency continue to be difficult to study because of the honey bee's open, uncontrolled foraging habits. In the recent past, this has also not been a funding priority in research. This field has more recently become quite popular, however, with the new emphasis on bee health in both the beekeeping and research communities.

Managing honey bee nutrition is one of the beekeeper's most important jobs. Supplementing colony nutrition need is, as beekeeping itself, more art than science. The food generally has to be prepared from ingredients purchased elsewhere. It then has to be administered to the colony at the correct time and under optimal conditions to be sure it is consumed.

Feeding Carbohydrates

Monitoring carbohydrates in a colony is relatively easy. It is generally done by estimating the number of frames of nectar or honey. Another way is to heft the colony — get an estimate of its weight in order to make a decision. Perhaps the best food is combs of honey or nectar. Beekeepers routinely shift these among colonies to ensure all have adequate resources.

The traditional bee feed is cane sugar crystals mixed with water. The resultant sugar syrup is then administered as needed. You should mix the syrup in different strengths depending on conditions. The usual proportion is one part sugar mixed with one part water by weight.

The formula can be varied based on the time of year: a weaker solution in spring will be more nectar-like and stimulate greater population growth. In the inactive time of year, more sugar than water is advisable. This is especially important in colder weather, when it is more difficult for the bees to evaporate excess water.

Cane sugar used to make syrup must not have impurities. Processed white sugar is best. Be careful not to use anything off color, tinged with molasses, or otherwise contaminated. Avoid other sugar sources such as maple syrup, and candy and soft drink manufacturing by-products.

A bewildering variety of feeders is available from commercial outlets or has been developed by enterprising beekeepers. Again, the beginner should stick to the standard types described in chapter 5: the Boardman, division board, or top feeder.

In the 1970s, another product became available to beekeepers based on plentiful corn production in the United States. High-fructose corn syrup (HFCS) was a great advantage, particularly to large-scale beekeepers. It required no mixing and came in two concentrations, 42 and 55 percent solids. The latter product is almost exclusively used at the present time. Some beekeepers realized early on that HFCS did not stimulate bees to produce population to the same degree as sugar syrup. They alternated them, therefore, depending on use; cane sugar syrup for population growth, HFCS for maintenance.

Use only pure cane sugar syrup in the feeder.

Smaller beekeepers did not have access to HFCS because it was sold in bulk, although some suppliers sell it in small quantities. Purity is also an issue with this product and some has been found to be contaminated and/or to contain high levels of a substance known to be toxic to honey bees, **hydroxymethylfurfural** (HMF). To be on the safe side, beginning beekeepers should stick to pure cane sugar syrup.

Use of HFCS also brought into focus the fact that any syrup fed to bees in bulk can be processed by bees, just like nectar, and turned into a honey-like product. This practice results in adulterated honey, which cannot be sold as pure honey in the market. Beekeepers, therefore, should refrain from feeding large amounts of syrup to colonies that have honey supers installed.

In a true emergency, colonies can be fed dry sugar crystals, which are sprinkled either on the bottom board or inner cover. The bees must be able to get to the sugar and a source of water to dissolve it for their use, impossible if the weather is too cold.

Feeding Protein

Monitoring protein in a colony of honey bees is difficult. There is no good way to determine how much is present or the amount to be fed. The easiest way is to see how much jelly the nurses are feeding to second-instar larvae. Another possibility is to consider any colony with drone brood to have adequate protein on hand. At any sign of nutritional stress, drone production is shut down and adult males may be evicted from the colony.

The situation is also complicated by the fact that two kinds of materials have been described in the literature, pollen substitute and pollen supplement. The former suggests that it is a complete substitute for pollen; the latter has pollen added. Bees prefer food with pollen added, which contains attractive elements important for consumption. No substitute has been found for these "phagostimulants." In reality, most substitutes must be considered supplements because generally bees bring at least some pollen in from the field while they consume the substitute.

Honey bees begin sampling a nutritious pollen patty (see pages 128–129).

Cooking for the Bees

If you want to create your own nutritional supplements for your bees, here are a few simple recipes.

Bee Candy

Although liquid syrup is generally fed, you can also feed cane sugar as candy, especially in temperate areas with harsh winters. The recipes generally use table sugar and water. Dick Bonney's recipe is as follows.

1. Bring 2 quarts of water to a boil in a medium to large pot. Turn off the heat and pour in a 5-pound sack of sugar, stirring until it is dissolved.
2. Turn on the heat, bring to a boil, and continue heating until the mixture reaches the hard-ball candy stage at 260 to 270°F. A candy thermometer is a must. The boiling process may take 30 to 40 minutes.
3. When it has reached the hard-ball stage (a teaspoonful turns into a ball when dropped into room temperature tap water), pour the mixture on sheets of wax paper on a flat, hard surface. If working on a finished surface, insulate it with plenty of newspaper under the wax paper, so it doesn't get damaged. Raise the edges of the wax paper with frame bars or similar-sized sticks to keep the candy from running off.
4. When set, the candy will be hard, somewhat brittle, and a light amber. Pieces may then be laid on top of the inner cover or directly on top of the hive frames. The closer to the cluster, the better it will be consumed.

Alternatively, the hot mixture can be simply poured into a spare inner cover, rim up, or even a modified outer cover, where the candy will set. This "candy board" is then inverted over the nest. Bees use the water that evaporates off the cluster to help them consume the candy.

Making Sugar Syrup

As temperatures drop, bees are less able to evaporate excess water; in fall and winter, therefore, thicker, more concentrated solutions are used than in late spring and summer. In general:

- **To install packages** and/or to stimulate brood or comb production: one-to-two ratio (by weight or volume); one-third sugar to two-thirds water; about 1.5 pounds of sugar per gallon of water.
- **Spring or summer feeding** to sustain colonies low on food or administer medication: One-to-one ratio (by weight); one half sugar to one half water; about 3 pounds of sugar per gallon of water.
- **Fall feeding** for colonies low on stores or to administer medication: two-to-one ratio (by weight); two-thirds sugar to one-third water; about 5 pounds of sugar per gallon of water.

Use only crystallized pure cane sugar.

Pollen Patties

Pollen substitute or supplement is usually fed in patties. Here are two common recipes.

For a metric conversion chart, see page 206.

Pollen Substitute

- Mix 1 part dry mix (1:3 pollen/expeller-processed soybean flour) and 2 parts heavy syrup (3:1 sugar/water).
- Add sugar to water and stir until dissolved.
- Add dry ingredients.
- Mix thoroughly and separate into 1.5-pound patties.

Soy-Sugar Patties

Another type of patty consists of 1 part soybean flour and 2 parts heavy sugar syrup, or 1 part dry mix (4 parts soybean flour:1 part brewer's yeast) and 2 parts heavy sugar syrup.

A 10 percent natural pollen mix can be added to any of the above recipes to make a supplement. (Beekeepers should be aware, however, that feeding pollen from a diseased colony can transmit diseases.)

Tips for Supplements

The soybean flour called for above must be less than 7 percent fat. Because of this, typical recipes call for "expeller-processed" flour, not solvent-extracted flour, which might inject impurities. Other materials have been used besides soy flour, including brewer's yeast, torula yeast, Wheast, canola flour, linseed flour, and sunflower flour. Still other ingredients have included baker's yeast, vitamin and mineral supplements, fish meal, peanut flour, skim milk, powdered egg yolk, powdered casein, sodium caseinate, and lactalbumin.

A yeast-based pollen supplement consists of 3 parts brewer's yeast and 1 part natural pollen added to 2 parts heavy sugar syrup (6 parts sugar:1 part water). There are also a number of commercially prepared protein patties sold on the market. A new liquid-based diet is expected to be available in the near future.

Take a quantity of dry pollen supplement or commercially available substitute and mix it with enough sugar syrup to make a dough. Form this into thin patties, place them between wax paper sheets, and lay them on the frame tops. On top of the brood frames usually works well, but don't place them in such a way as to break up the brood nest.

Pollen patties dry out in the hive. The amount of sugar syrup used in mixing is a balance between making the dough stiff enough to hold shape, but damp enough to remain usable. They can become rock hard. Remove and replace those that stiffen up. Smaller ones fed more often are helpful in controlling this problem.

Nowadays top-tier commercial pollen substitutes ("sub") are nearly as nutritious as natural pollen. You get what you pay for.

Be sure that pollen patties are placed in full contact with the cluster, because bees will not consume sub placed outside it.

The feeding of pollen sub during periods of pollen dearth is one of the most useful management tools for the beekeeper. It can rejuvenate a colony.

8

TAKING THE CROP

Harvesting the honey crop is a chore that looms large for many beekeepers. It often seems to be such an overwhelming task that it simply does not get done the first year or two of keeping bees. Novices sometimes leave honey supers in place the first winter rather than face the problem. This can, of course, lead to a further problem in spring if they find that the colony has moved up into the honey supers over winter and is happily raising brood there.

The Honey Crop

Most beginning beekeepers have one goal: produce as much honey as possible. The bees usually make more than they can use. The beekeeper judges the amount the bees need and harvests the surplus. A rule of thumb is never to take more honey than the colony will need to survive the rest of the year, including winter. This is easier said than done. Beyond quantity, other hard harvest decisions have to be made, including when to remove honey, how, and under what conditions.

Honey is infinitely variable, which you can easily see if you examine (in a bright light) comb filled with honey from different nectar sources. The range of honey products available in stores and farmers' markets also attests to this.

The official definition of honey is continually under review to protect the consumer from purchasing an adulterated product. Here is the current definition, devised by *Webster's New World College Dictionary* and approved by the US Food and Drug Administration:

> *A thick, sweet, syrupy substance that bees make as food from the nectar of flowers and store in honeycombs.*

The best way to identify the plant source of any honey uses pollen analysis. Honey that is filtered to remove pollen cannot be identified with any degree of certainty.

The color of honey ranges from dark amber to water white. The final color is based on the floral source of the nectar. Tropical honey tends to be darker than that found in temperate regions, but not always. Some beekeepers try to separate darker from lighter honey, but it is easiest for the beginner to mix the crop together at the start. The resulting honey is likely to be darker than that found in the store. Although some customers may question the quality of this product, most find that darker, stronger honey has more character. Increasingly, people appreciate the darker product for its higher proportion of antioxidants, vitamins, and minerals.

Liquid gold pours from a honey extractor.

HONEY: LIGHT OR DARK?

Many consumers have a conditioned preference for lighter honey, which generally commands a premium price. As a consequence, many beekeepers are prejudiced against the darker and sometimes stronger-tasting honey from the late season, considering it good only for winter stores or for sale as bakery-grade honey. This is changing, fortunately. There are very few honeys that do not taste good, but honey does come in many flavors. Increasingly, people realize that all honey is not the same.

Honey Harvesting Q&A

Common questions about harvesting focus on the how, the why, and the when. The answers are interrelated and some have been discussed elsewhere in this book, but it makes sense to consider them again in the context of taking the crop.

When do you take off honey?

Do you wait until the end of the season, or do you start harvesting honey as soon as there is a full super? The answer to this question depends on the nectar flows in your area, your work schedule, and your honey preferences. In some areas, the main nectar flow is over by midsummer, with nothing significant beyond that. In other areas, there is a somewhat continuous nectar flow through most of the season. In still other regions, a main nectar flow occurs in the early summer, followed by another significant flow in the late season, with the honey from the early flow usually being lighter and milder.

The first question you must answer is: What is the nature of the nectar flows in your area? A second question: Do you have preferences toward any of the seasonal varieties? Here are the choices of action that follow.

If you have more or less continuous nectar flows throughout the season you may choose to wait and take off the honey at the end, or you may take it at intervals throughout the season. If you are able to separate varietal flows (from different flowers), this may be desirable. If you have large crops, you may wish to spread the work over more than one extraction and you may wish to minimize the number of supers you own by reusing them during the season.

If you have a single early flow, you will normally take off honey once, at the end of your season.

If you have both an early and a late flow, there are two approaches: take off honey after each flow, or wait and take it only after the final flow. The former approach will give you two separate crops, probably of different color and taste, while the latter will give a blend. The former will also be more laborious while the latter allows you to own fewer supers.

A basic rule of thumb is: do not take the honey until you are sure that the colony does not need it for winter. This can be overridden, of course, if you are prepared to feed to compensate for taking those potential winter stores, but it is a questionable approach that may do more harm than good.

How much honey should you take off?

Simply stated, you take off only that which is surplus to the needs of the colony. To establish what these needs are, ask yourself these questions:

- Is this the end of the main nectar flow?
- Are there minor flows to come?
- How long is it until winter begins?
- How much will the colony consume between now and then?
- How much will they need to make it through the winter?
- Although there may be honey in the supers, is there also the requisite amount in the hive bodies?

Your goal is to come to the end of the season with as much honey or stores in the hive bodies as is required for overwintering in your area. Precise amounts are difficult to state here. Requirements vary throughout the country. The practices of beekeepers in your area, modified by your experience, are your guide.

Is it all right to leave extra honey supers on for the winter?

Although there are exceptions, generally it is best not to leave extra supers with honey on the hive during winter. Two deep hive bodies will suffice for winter quarters and honey storage in the North, as will a single deep with a "food chamber" shallow in the South.

If you are leaving extra supers on because otherwise there may not be enough stores for winter, this means that the hive bodies have not been efficiently utilized. They are not full. Even though leaving a super may give them enough food reserves, the food may not be well distributed in the hive. The bees may become isolated from the honey if it is not efficiently stored, especially in very cold weather when movement in the colony is restricted.

A second problem is that in spring the colony is most likely to have worked its way up and is living and raising brood in the super. You then face the problem of clearing the super so that it can be returned to its normal function. Although not a major problem, this is at least a nuisance. To clear the super, place it under the bottom hive body in spring. The colony will normally move up out of it to the hive body above as the brood matures. The super, of course, will suffer some minor wear and tear, and the comb will have darkened, from being used to raise brood.

Should you remove the queen excluder for winter?

There are occasions when you may decide to leave a full super on the hive for winter. Even so, you should remove the queen excluder. Otherwise, the colony, or at least the bulk of the colony, may move up through the excluder as the season progresses, with the danger of leaving the queen behind. At worst, she could be isolated and die; at minimum, she would not be able to develop an efficient brood nest as the new year begins.

What do you do with unripe honey in the supers?

First, be sure that the honey is, in fact, unripe, and not simply uncapped. It is not uncommon in the late season for the bees to leave a larger than normal amount of honey uncapped. To make this determination, look at the honey: cells of unripe honey usually will be little more than half full. Hold the frame horizontally and shake it; unripe honey will drip out.

Stick your finger in and get a taste; unripe honey will taste green, closer to flower nectar.

If you determine that you are dealing with unripe honey, feed it back to the bees (see box). It will ripen before winter.

What if you have some honey in the supers but not enough in the hive bodies for wintering?

The same technique of feeding back applies here as for unripe honey (see box). Scratch the cappings, put the supers over the inner cover, and let the bees move it. If there is more honey in the supers than the bees need, you can extract it first and then feed back what they need in a conventional feeder.

FEEDING HONEY BACK TO THE BEES

Feed back the honey by putting the super over the inner cover in the late season, first scratching any cappings that do exist. Your goal is to do a little damage (only a little) to the comb and cappings so that the honey will ooze and run a bit. This will prompt the bees to go up through the hole in the inner cover and remove the honey, bringing it down and storing it in the hive bodies.

This must be done in the late season, as the nectar flow is winding down. Done too early, the bees will not move the honey. They may ignore it completely or they may repair any damage that you have done and continue to use the super, even though it is above the inner cover.

Harvesting the Crop

Over time, the beekeeper becomes attuned to local conditions and will be able to estimate when to take the crop and how much to leave for the bees. Some areas will have a continuous flow; others will be stop and start. Late honey flows can often be as large as, or larger than, earlier ones. Certain populations of bees will consume more during inactive times than others. Both bees and beekeepers can and will make mistakes, so the human manager must stand ready to step in if too much food is removed too soon. Remember that honey is the best material to leave in the hive for the colony's survival.

A key factor in removing honey is moisture content. Generally, honey must be at or below 18.6 percent water before it is "ripe" enough to harvest.

When bees determine the moisture content is correct, they cap it over with comb wax. Thus, another rule of thumb is to remove only honey from frames that are two-thirds capped. As in other aspects of beekeeping, it is much better to err on the side of harvesting too little than too much. Honey will absorb moisture from the air during processing, so it is important to be conservative here — process honey only after it has been capped, and limit its exposure to warm, moisture-laden air during extraction and bottling.

There are four common methods by which to remove honey bees from supers: brush, escape board, fume board, and blower. Each has its own set of challenges. No matter which method you use, be sure to have some empty supers available in case you need a place to put the frames after the bees have been removed.

All four of the techniques described below can be supplemented with brushing if many

BREAK BURR COMB AHEAD OF TIME

In spite of the bee space, the bees don't always leave certain gaps wax free. Partial comb buildup is usually found on top bars and around the queen excluder. This burr comb may contain honey and brood and interfere with super removal.

Break off burr comb and/or scrape it off with a hive tool before you remove the honey. Work quickly to prevent robbing and then put the colony back together, giving the bees a day or two to clean up whatever mess is left. You should then be able to remove the supers relatively easily.

Burr comb can build up on top bars. Remove it or break any connections between hive bodies to facilitate removal of the supers.

bees remain, but this makes a one-step operation into two. Make sure a brush is always available, however, no matter which method you choose.

Brushing

Brushing is easy, inexpensive, and relatively quick. If you are harvesting from many supers, however, brushing becomes tedious, and it's difficult to keep the bees from returning to the frames even as they are being brushed. Brushing also can provoke robbing.

Before you begin, place an empty hive box on a solid bottom board a little distance from the colony, where you can put the brushed combs one by one after the bees are removed. Cover it with a damp towel. If you don't have an extra super, improvise with a cardboard box or other container, but be sure bees can't get access to the just-brushed combs.

Several kinds of bee brushes are available. Most beekeepers prefer those with plastic bristles, but these are harder on bees than softer ones made of natural bristles.

Time is of the essence to prevent robbing of the honey by bees from other colonies. Move the brushed combs inside a hive body or box as soon as possible.

THE "BRUSH-OFF"

1. Remove the comb and hold it over the hive entrance. Shake as many bees off as possible.
2. Holding the frame over the open hive, quickly, gently, but firmly brush the others off with a sweeping motion. It is impossible to get all the bees off; don't waste time trying, just remove the majority.
3. When the comb is mostly free of bees, transfer it to the empty box or container.

Using a Bee Escape

The bee escape is an elegant piece of equipment that acts as a one-way valve. Bees can exit through a set of flexible springs but cannot return. The inner cover has an oblong hole that accommodates commercially available (sometimes called "Porter") bee escapes.

Alternatively, you can construct special escape boards. Be sure there are no cracks or other ways robbers might enter above the bee escape. Duct tape can be used to seal any areas of concern. With the escape board in place overnight, the vast majority of bees should leave the super, and you can then remove it as a unit, fairly free of bees.

Before using this method, check the combs in the super you are harvesting to make sure it is free of brood. The presence of brood above the escape board reduces efficiency because the bees are reluctant to leave the developing offspring.

Most large-scale beekeepers use a **fume board** to drive the bees out of supers. A strong-smelling liquid, such as benzaldehyde (oil of almond) or butyric anhydride, is dripped onto the board, which is then inverted over the colony. Because of the fumes, most of the bees leave the supers in a very few minutes, a big advantage over other methods described here, which take much more time. Again, as with all methods, it is impossible to remove all bees.

This technique varies in effectiveness, depending on weather and other conditions. Too little evaporation of the fuming material means many bees will remain in the super. Too much evaporation during high temperatures drives bees entirely out of the brood nest, which is not the intention here.

Beekeepers have tried many kinds of liquids over the years, some of which are no longer legal. The beginner should not experiment; buy a labeled product from a beekeeping supplier and follow the directions scrupulously.

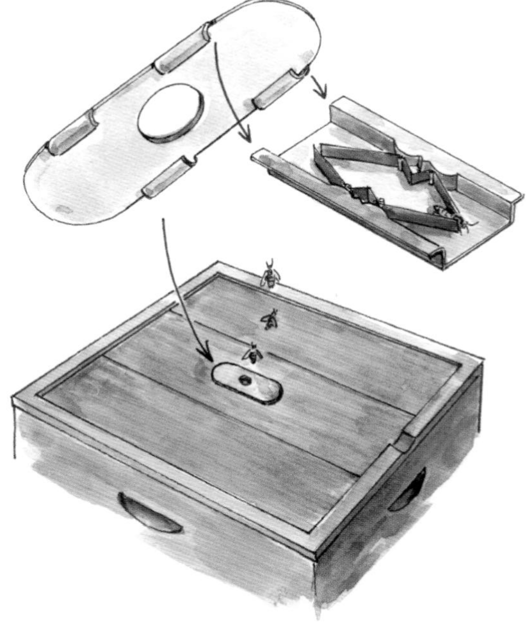

A bee escape has springs that act as a valve allowing a bee to exit but not return. The oblong apparatus fits in the inner cover.

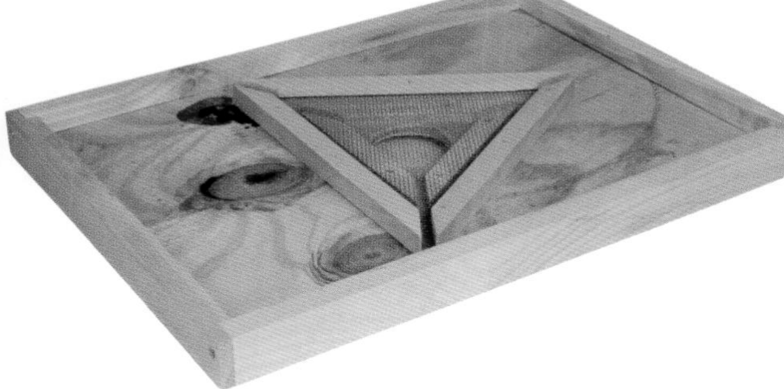

Designs for elaborate "escape boards" are prevalent in the literature and some are available commercially.

USING A FUME BOARD: Volatile chemicals absorbed into the top of a special cover or fume board drive the bees down into the colony, clearing the honey supers.

Clearing bees with a **bee blower** requires a major investment in either purchasing or making a suitable apparatus. Many commercial leaf blowers are suitable, but not all. Some beekeepers have modified vacuum cleaners, and many other designs have been developed over the years and published in the literature.

To use this method, remove the super and place it on some kind of stand in front of the colony. Direct the air between the frames. The bees blown out of the super and off the combs will quickly return to their parent colony. Too much burr comb reduces this method's effectiveness. You will be surprised how tenaciously some bees will hold on to the comb.

USING A BEE BLOWER: Beekeepers remove bees from the super by blowing air between the frames.

THE SMALL HIVE BEETLE

In the past, beekeepers could remove honey, store it for extended periods in the comb, and extract the crop in a leisurely fashion. This is no longer the case in many regions. The small hive beetle (*Aethina tumida*), first detected in 1998 in the United States, will lay eggs in stored honey. The resulting larvae add microorganisms that change honey's consistency and chemistry. When this happens, the sugars begin to ferment and the honey becomes "slimed." It is then not suitable for human consumption, and the bees also ignore it.

The small hive beetle is a tropical organism, and so its activities are regulated by temperature. It is, therefore, much more of a problem in warmer than cooler areas. The larvae seek light because they must exit the hive to pupate in the soil.

Beekeepers can shine a strong light on the floor of their extracting facility to attract the migrating larvae out of the honey-filled supers. Keep a bucket filled with water nearby to drown the wormlike larvae.

The conclusion is clear. In regions where the small hive beetle exists, beekeepers no longer have the luxury of extracting the crop at their leisure. Supers should be removed and immediately extracted and comb honey frozen to prevent damage from this destructive insect. (See pages 184–186 for more on the small hive beetle.)

Processing the Crop

Processing the crop usually consists of two separate activities: removing honey from the comb (extracting) and then storing it in a suitable container. Both are vitally important if the resulting product is to be of a quality that consumers will want. A general rule is the less processing the better, which is the reason a growing number of beekeepers are concentrating on marketing honey in the comb.

Removing Honey from the Comb

Extracting requires first that you remove the cappings and then remove the honey from the cells. For the beginner, two strategies exist to remove honey from the comb: crushing it, caps and all, and straining the honey; or using an extractor. The latter requires that you first remove the cappings so the honey can be literally slung out of the comb.

Straining

The simplest way to crush honey comb is to put it in a nylon stocking and then squeeze it, forcing the honey to separate from the comb. Squeezing and forcing take time and can waste a lot of honey.

Honey is **hygroscopic**, which means it absorbs water from the air. Warm honey flows better and can be processed quickly, but this can also result in more moisture in the product. Both ambient humidity and temperature, therefore, are important considerations if you use the straining method because the honey is likely to absorb a lot of water in the process. A dehumidifier can mitigate this situation to some degree.

The disadvantages of straining are the relatively long time it takes — exposing the honey to more and more moisture — and the fact that the comb is destroyed and cannot be reused. If you have to harvest a large amount of honey, crushing the comb and straining are simply not practical.

Use a strainer to filter honey coming out of the extractor.

Using an Extractor

Two kinds of extractors exist: radial and tangential. Both spin uncapped frames at high speeds, literally slinging the honey out to collect first on the extractor walls, then drip to the bottom of the holding tank. Many models are available, holding from 2 to 60 or more frames. Some extractors spin the supers themselves, frames, hive boxes, and all.

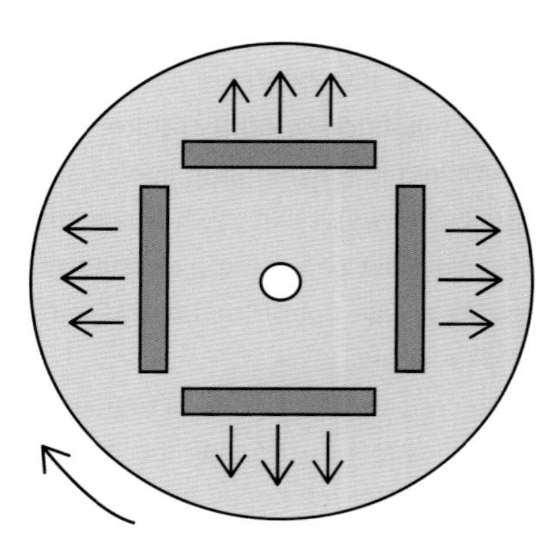

tangential

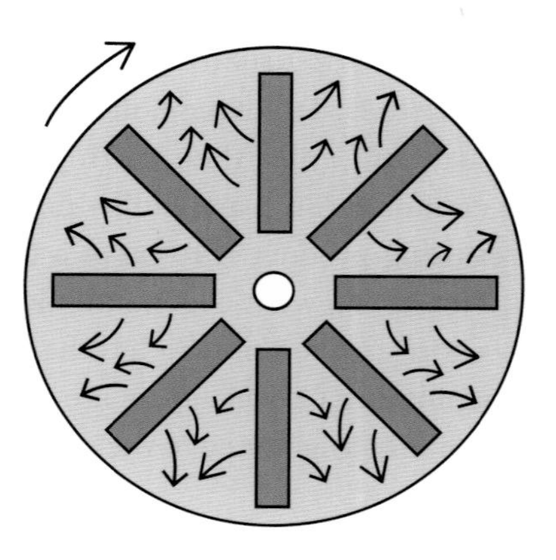

radial

Honey is slung out of the comb using a tangential extractor.

To save time and avoid muscle strain, you can attach your own simple electric motor to power the extractor.

An extractor is expensive, but there are many alternatives to purchasing one. Some associations lend extractors or even extract for their members. So-called custom extractors are also sometimes available to do the job for a fee. Large-scale extracting requires other equipment, including automatic uncappers, honey sump pumps, holding tanks, honey gates (valves), filters, and heaters.

Storing the Crop

Honey can be stored for a long period. The sweet from a hive is, like wine, a "living" organism filled with enzymes, proteins, and sugars, all capable of changing the honey's properties over time. Some apocryphal stories have been published about honey found in ancient amphoras still being edible.

Avoid Excess Moisture

Too much moisture will destroy honey. Once the moisture level rises to more than 19 percent, natural yeasts present in honey begin to ferment, producing alcohol. If honey is allowed to stand for a long period of time, it might crystallize. This process separates out the solids and thereby increases the liquid portion, promoting fermentation.

Many consumers believe crystallization ruins honey, but that is not the case. Only fermentation does. If a distinct alcoholic odor is not discernible, fermentation has not begun. You can easily reliquefy crystallized product by immersing the container in warm water.

Control Temperature

Temperature can affect honey in several ways. Most commercial honey has been flash heated to 160°F (71°C). The sweet can be brought to 120°F (50°C) for easiest liquefaction and filtration, but should not remain at this temperature for long. Warm temperatures denature enzymes like diastase in honey and also will convert the fructose in the sweet into hydroxymethyfurfural (HMF), which at high levels is toxic to bees.

Honey can be frozen with no ill effects. The optimal crystallization temperature is 57°F (14°C). Store it above or below that temperature for best results.

Honey in bulk is slower to granulate than when placed in smaller jars. When storing honey for extended periods, therefore, it is better to use larger containers. Smaller beekeepers will use 60-pound or 5-gallon square tins or 5-gallon plastic food-grade buckets with lids. Others may have large holding tanks containing up to several 600-pound barrels. You can bottle the honey in smaller quantities over time when you wish to sell or consume it.

Chunk and Comb Honey

Many people like to consume honey in the comb. For this purpose, beekeepers often use what is called **cut comb**: they cut a thin piece of comb from a frame and cover it with strained honey in a jar. Generally marketed as "chunk honey," this is an excellent way for a beekeeper with limited means to process the crop, as it reduces the amount of extraction needed.

Before modern food inspection rules and laws, processors would often intentionally adulterate honey by adding a number of substances. Since all liquid honey was suspect, the best way to thwart adulterers was to sell it strictly in the comb. Honey in small frames or "sections," therefore, was sold all over the country from 1880 to 1915, when train-car loads of the product were shipped to urban markets.

Producing quality section comb honey is not easy. It requires large, established populations of honey bees and strong honey flows. The insects do not like to work in the close quarters demanded of them in frames designed to hold comb honey sections. They must be forced to do so, and this can result in a lot of incomplete sections of questionable quality. In recent years, different kinds of technologies have been developed to ensure a better product with less labor. The production of comb honey is not generally recommended for the beginner, who should stick to chunk and/or extracted or strained liquid product.

HONEY AND HUMAN HEALTH

Although primarily thought of as a sweet food, honey has been used as a remedy and treatment for human ailments for centuries. Sore throats, coughs, and other conditions, including hangovers, are thought to be helped by the sweet. In addition, honey has been used as a topical dressing for burns, wounds, and other injuries. It contains an enzyme known as glucose oxidase that forms, as a by-product, hydrogen peroxide, which kills many kinds of bacteria. Some patients suffering from a pernicious form of bacterium, methicillin-resistant *Staphylococcus aureus* (MRSA), have seen remarkable results when treated with honey.

Some honey is more potent than others. Manuka honey, made by bees from the flowers of the New Zealand "tea tree," has known antibacterial properties. The Active Manuka Honey Association (AMHA) promotes the honey for medicinal uses through its unique manuka factor (UMF) that grades the antibacterial strength of each batch.

Forms of Comb Honey

Four forms of comb honey are commonly produced and marketed. The differences have to do primarily with the packaging. They are round comb, section comb, cut comb, and full frames.

Round comb is produced in 4-inch comb honey supers that contain plastic frames and fixtures. As it has no corners and holds less honey than square section comb, it is more quickly and readily filled by the bees.

Section comb is also produced in 4-inch supers. This old standby in its square wooden boxes has been around for many years.

Cut comb is usually produced in standard shallow extracting supers, or occasionally in mid-depth supers. Standard extracting frames are used, although without wires in the frames or the foundation. The completed full comb can be cut out of the frame in four equal 4¼-inch sections, and each section is packed in a square plastic box.

Full frames may be produced in any size super but are most common in shallower varieties, not full depth. The frame with comb can be marketed in a suitable box or wrapping. Somewhat unwieldy as a consumer item, these are most often seen as entries in honey shows.

Producing Comb Honey

Many successful comb honey producers reduce the colony to one hive body at the time they put comb honey supers in place. As much of the brood as possible is placed in that one hive body, which should be "boiling over" with bees. A queen excluder may or may not be placed over the hive body, depending on the beekeeper's thinking, and up to two comb supers are added.

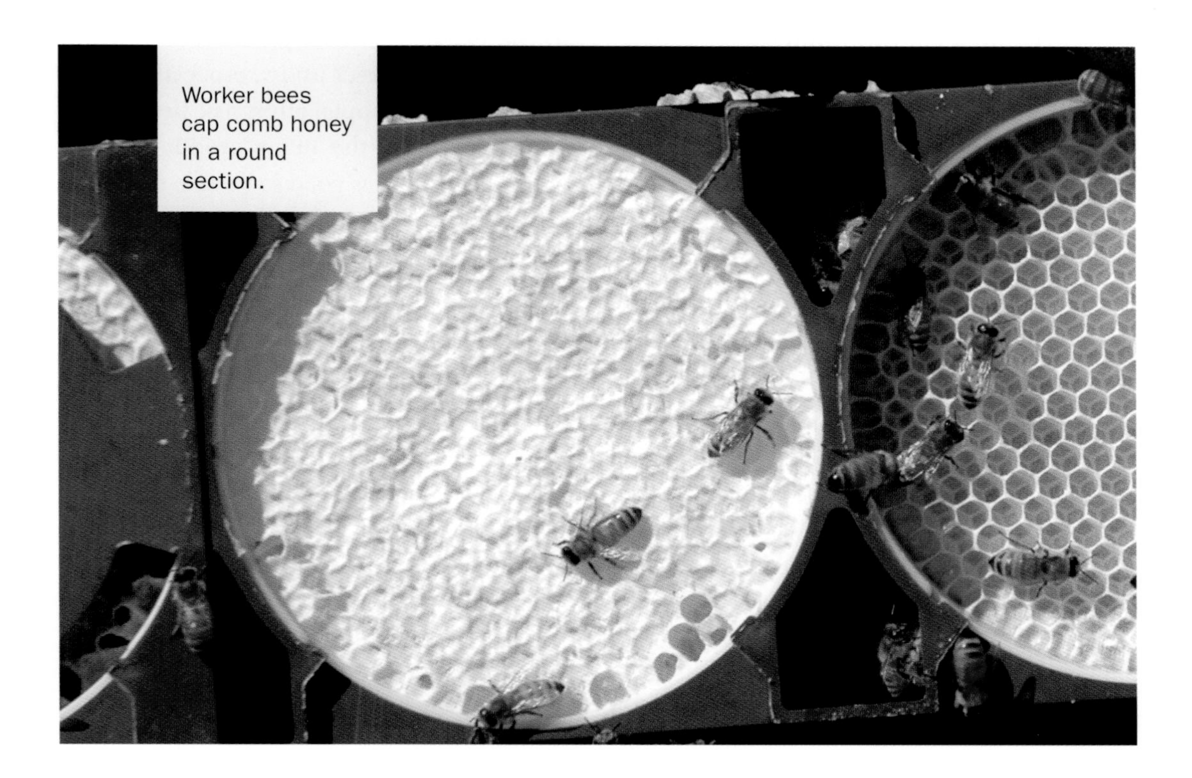

Worker bees cap comb honey in a round section.

Reducing the available space for brood rearing by squeezing the colony to one chamber shifts the bees' attention and creates the best environment for comb honey production, especially during an intense nectar flow.

A Suggested Method for the Novice

Even with limited experience, it is possible to produce a little comb honey, and sometimes even a large amount. Give it a try. Place a comb honey super with unwired cut-comb foundation on the hive. Do this when there is a good nectar flow in progress and the colony is boiling with bees.

If all goes well, the colony will draw the foundation and fill those frames with honey. Inspect frequently and remove the comb honey frames as soon as they are full and capped, substituting more comb foundation if the honey flow is going well. If not, you can replace them with wired extracting foundation.

Little is lost should the product not live up to expectations. If by the time the nectar flow is over, those comb honey frames are not fully drawn out, not totally filled with honey, or not completely capped, treat them as extracting frames. The only difference between these and regular wired frames is that they will not be as strong during extraction. In each succeeding year, however, the bees will strengthen these combs a little more.

A variation of this method is to use comb foundation that is wired in the frames, as is often done with regular brood or extracting frames. The wire can be removed at the time the comb honey is harvested by first cutting it on the outer side of the end bars and then drawing it through the holes in the end bars with a suitable pair of pliers. This is best done by first heating the wire quickly and briefly. Beekeepers usually do this with a device similar to an electric wire embedder.

Other Bee Products

Although honey is the main product available to the beekeeper, honey bees also produce other items of direct utility, especially pollen and beeswax. People often ask beekeepers questions about the many products of the honey beehive. It is good to be able to discuss them all in some detail, and to offer resources for more specific information.

TO YOUR HEALTH

Beeswax, pollen, propolis, royal jelly, venom, honeydew, and honey: all are marketed and sold as health supplements or natural foods. There is little direct scientific evidence to support many of the health claims for these products, but there is growing interest in their properties. The American Apitherapy Association is dedicated to providing information on the use of honey bee products in a branch of human medicine called apitherapy. This alternative therapy employs bee products in a variety of ways to treat human health problems ranging from allergies, wounds, and burns to chronic problems such as multiple sclerosis. In some countries where modern medicine is not available or too expensive, apitherapy is the basis for what has been called "green medicine." This is the case in Cuba, which has a significant body of physicians engaged in this activity.

The **solar wax melter** with a glass top concentrates heat, melting the wax in an upper tray. The wax then runs off and accumulates on the bottom of the apparatus.

Beeswax

Beeswax is unique and can be made only by honey bees using specialized glands. It is literally the foundation for the bee nest, becoming both the cradle for future bee generations and the food warehouse of the colony. The elegant six-sided cell form uses the least amount of material for maximum strength and has been copied for many human applications. A recent spectacular example is the construction of the James Webb Space Telescope, partial successor to the aging Hubble Space Telescope.

Most of the world's beeswax is recycled into foundation and given back to the bees to manufacture comb. It is also an ingredient in many cosmetics, creams, and salves, and a component of wood and metal polishes. It is a good dry lubricant and used by carpenters and others in woodworking applications. Of course, it makes a superior candle and continues to be used in religious applications.

During the year, the beginning beekeeper will accumulate wax from scrapings of the hive and frames. The wax cappings from the honey crop and the wax rendered from combs that are being renovated also provide good sources of wax.

The best way to collect the finest wax is to use a solar wax melter. Beekeeping literature contains many plans for building this kind of apparatus, which is usually nothing more than some kind of insulated box covered with glass and set in direct sunlight (see Resources).

Pollen

Pollen can be collected from a hive and later fed back to the bees or sold as a health food. Specialized pollen traps at the hive entrance force worker bees to clamber through small openings. In the process, any pollen balls located on the last pair of legs are literally scraped off the insects.

Because traps are hard on colonies, however, do not place them on those that are weak in population. Even on populous colonies, remove them periodically to give the bees a chance to replenish their protein reserves.

Pollen is fragile and quickly loses its nutritional value when exposed to the air. For use in bee feed, it is best to mix fresh pollen with granular sugar and freeze it until needed.

Dried pollen is sold to health food stores. It is often touted as a perfect food for humans, but there has been little research to support this claim. One reason is that all pollen is not the same and has radically different characteristics based on its plant source, just like honey. For bees, this means that a diet of mixed pollen is preferable to that from a single plant source. For human consumption, this means that a standard dose of pollen for various ailments and applications does not exist.

Pollen is also a magnet for environmental pollutants, whether particulates from smokestacks or pesticides. There is generally no way to know the source of collected pollen, and therefore, no easy way to ensure its purity for either bee or human use.

Royal Jelly

The thick, white food that workers secrete from their hypopharyngeal glands and feed to queen larvae is called royal jelly, thought to be responsible for the queen's special qualities. In fact, the jelly is the same as that fed to developing workers; the queen's genes are turned on by dining almost exclusively on the substance, whereas that fed to workers is modified by adding honey and pollen. The idea of food "fit for a queen," however, has taken hold of the human imagination. It is a healthy, nutritious food for bees. How humans benefit is not clear, and collecting it is an expensive way to get only a relatively small amount of nutrition.

Propolis

Propolis is a resinous material collected from the bark and buds of trees and other plants. It is the "bee glue" of a colony, used to plug up small holes and cracks. In some cases, bees use it to cover unwanted objects in the hive that they physically cannot remove. The classic example is a decomposing mouse carcass on the bottom board.

Recent research is providing better information on how propolis helps bees ward off diseases. Research in human medicine has shown propolis to be an antimicrobial agent, an emollient, immunomodulator, dental anti-plaque agent, and anti-tumor growth agent.

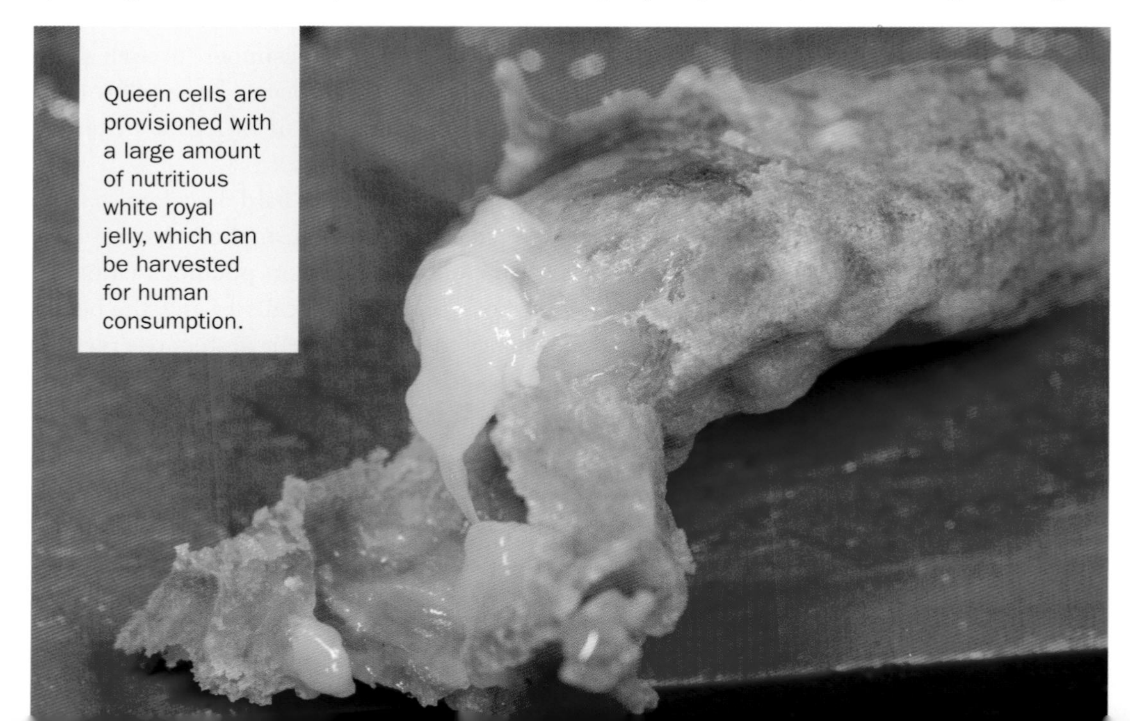

Queen cells are provisioned with a large amount of nutritious white royal jelly, which can be harvested for human consumption.

A close-up of propolis, the "bee glue" of a colony

Studies also indicate that it may be effective in treating skin burns.

As with nectar and pollen, however, propolis is infinitely variable, so the propolis from one region is unlike the propolis from other areas that have a different plant mix. Finally, it should be noted that in the absence of natural propolis, bees may collect materials that are analogous, such as paint, road asphalt, and other objectionable substances.

Venom

Honey bee venom is a complex of many different chemical substances. It is possible to collect it directly from bees using special devices, and research has revealed therapeutic effects in humans, especially in reducing symptoms of some forms of arthritis. At least one producer sells honey with venom added. Bee venom is also used to desensitize people who are allergic to bee stings. This technique has been part of mainstream medicine for decades.

Because it is produced by bees and not collected from the environment, venom seems to be a uniform product from which a standardized dose can be formulated. Collected venom, however, could be significantly different from that obtained directly from a honey bee through a sting because it has lost many of its volatile components. This is why most apitherapy practitioners use live bees. Several have developed procedures to prevent bees from losing their sting apparatus during collection, although it is not clear whether individual bees can regenerate venom or are accepted back into their parent colony.

Honeydew

Honey bees also produce a variant of honey called **honeydew**. This is a sweet material collected and processed just like honey, but its source is other insects, not flowers.

In certain parts of the world, forests support large populations of plant-sucking insects that literally plug into the circulatory system of a plant by drilling through the bark. They have to suck a lot of juice out of a plant's vascular system to get the appropriate nutrition, and in the process excrete surplus liquid in the form of a sweet juice, which honey bees then collect and modify in a manner similar to making honey.

The resultant honeydew is often in great demand in local markets. It generally has a markedly different flavor than honey. Major areas of honeydew production include pine forests of Germany, western Turkey and nearby Greece, and beech forests of New Zealand. Other regions, such as the Panhandle of Florida, also have honeydew flows that support the honey bee population, but are not sufficient to provide the beekeeper with a marketable product.

BEEKEEPER'S STORY

I LIVE CLOSE TO THE NORTH SHORE of Lake Superior. The climate makes beekeeping difficult, since the winters are long and cold. Spring is slow in coming because of the lake effect. Almost everyone winters in two deeps, after feeding 6 gallons of sugar syrup in October and leaving all the honey in the brood chambers in September.

I wrap my colonies in 2-inch-thick blue Styrofoam, with an upper entrance, and let them be buried in the snow; I just have to watch so the entrance doesn't ice over in spring. I have very good success wintering the hives, only occasionally losing one if a queen fails in spring, since we can't bring in replacement queens if we want to keep the mites out.

I start feeding again in spring, after digging the hives out of the snow, about the first week in March, giving each hive about 1 quart of 2:1 sugar syrup every week. When I see that the bees are short of pollen, I give them a pollen substitute.

This summer has been terrible so far. Nothing but cold and wet, and not more than three days when the temperature hit 80°F (26.7°C). Honey will be dear this year, if it doesn't warm up quickly in August.

Growing up in southern Minnesota, our family did not keep bees, but we ate a lot of clover and basswood honey. My dad had a beekeeper place hives on our property for pollination of raspberries for many years, and after I started with bees, I enticed my mother to get a few hives. She bought them locally, and I swear they were Africanized, even back in the '60s. They hung out of the hive for days after we worked them, stung everyone and everything in sight, and chased people through the cornfield.

I really didn't appreciate how ornery they were until I came home from Ohio State, where I was then working with Dr. Walter Rothenbuhler on the behavior and genetics of resistance to American foulbrood, and went out to examine Mother's bees in a T-shirt. I told her she probably didn't give them enough smoke, but no amount of smoking helped — they were horrendous. So I gave all of her queens an alcohol bath, got on the phone, ordered Buckfast queens, and put them in the hives. I left them in for a whole week, successfully introducing them. In 30 days, her bees were gentle as lambs, and they stayed that way.

One of my first experiences at university was learning what happens when a person opens a big colony too late in the day. I had caged a couple of queens in the brood nest of two big colonies, and after going fishing with my brother in the late afternoon, returned in the evening to retrieve my queens. I opened the colonies just as dusk was setting, and the bees boiled out of those hives by the millions!

I had blue jeans on and long sleeves and socks up over my pants, but the bees just crawled all over me, stinging as they went. I quickly closed the hives, since getting robbing started would have resulted in sure termination of employment, I figured. The next two days were spent in bed, 100 stings on each leg, and feet and legs so swollen I couldn't get shoes on. I learned that you don't open hives, even with a smoker, that late in the day.

When I started beekeeping, the biggest problem was American foulbrood. Breeding for hygienic behavior has largely solved that problem, at least in this area, as long as people put the effort into selective breeding. We don't look just for hygienic behavior; naturally, the colony has to make it through winter first. Second, they have to be nice to me, or the alcohol bath takes place. They have to produce honey, and then we breed from those that do those things, and are also hygienic. Not every queen will be hygienic, as the behavior is the result of two pairs of recessive genes on different pairs of chromosomes. But with selection over the years, we are doing quite well.

If the clovers and alfalfa are allowed to bloom, the bees, of course, use them. But farmers cut everything before it blooms these days, so whereas July used to be the best month for honey production, now it is about the worst. Spring is better, with willows, alder, pin cherry and moose maple, dandelions (of course), and the odd apple tree. We have wild red raspberries in July and August, followed by fireweed, goldenrod, and asters. The honey produced here is delicious — real wildflower honey — since so much clover is cut. Even the roadsides with sweet clover and trefoil are cut as soon as the poor flowers start to bloom. It is a wonder bees can survive at all. There used to be a few thistles around, but there are not many now. Every field is sprayed — not a weed in sight in any grain field, even.

I raise my own queens, and haven't brought any in since 1987 when the border with the United States was closed, trying to keep out the mites. Our district is still mite free, thanks to our bee association, and the beady eyes of a couple of bee inspectors who found one shipment that came into town with a few mites. We destroyed those bees before the mites could spread, and replaced the bees for the keeper out of our own pockets.

In the days before mites, I used to buy a few Buckfast queens each year. But even then I raised most of my own queens. The Buckfast stock we have here was brought in before the African genes were added to the line, as I understand it. There were also Carniolans before I arrived here, and some Carniolans were probably brought in from Hawaii as well, so we have kind of a mix. I try to raise a replacement queen from each colony each year, to try to keep as much genetic diversity as possible. Unfortunately, I am growing old and cannot handle very many colonies now, but I have a few young folks who are starting out, and I have them raising bees as well.

I sell some honey, although I donate a large amount to our local symphony orchestra. I also sell beeswax candles and hand cream, and a few queens and nucs. I sell the bees primarily to local beekeepers, as we still have no mites, small hive beetles, or Africanized honey bees in Thunder Bay District — and we all want to keep it that way! Beekeeping is still fun here, since we don't need to use all sorts of chemicals to control either bee diseases or parasites. Yet.

Before you get bees, read a few books about beginning with bees. If possible, take a short course on beekeeping. Join a bee association. Offer to assist an experienced beekeeper, maybe even for a year, to see whether you really do like beekeeping — and getting stung. You will get stung, and so will your significant others, most likely. So before spending too much money and time on this hobby, make sure you know something about bees and actually like working with them. If you do, let me warn you: it can be very addictive! I have been trying to quit for about 10 years now.

Jeanette Momot, Thunder Bay, Ontario, Canada

9

POLLINATION

The service that honey bees provide to plants far outweighs the collective benefits of their products, such as honey and beeswax, to humans. Yet all too often farmers and gardeners have downplayed the need for adequate pollination in comparison to other agricultural inputs influencing soil fertility, moisture availability, and disease and pest pressures.

Today's increased understanding that pollination is critical to plant success and food production is fueling a renaissance in the beekeeping craft. For some beekeepers, the pollination benefit will remain invisible as they focus on marketing the various products the honey bee offers. Many, however, are now shifting direction to provide the essential service of pollination, whether to increase production in a home garden or to develop commercial opportunities in the larger grower community.

Unique Challenges

Commercial pollination is a niche activity and completely different from processing and marketing products of the honey bee colony. It is a service with a set of unique challenges. Anyone considering this kind of enterprise must be ready to move colonies at a moment's notice into or out of fields to pollinate plants more efficiently and/or to protect colonies from damage by insecticide application. Because there are a great many opportunities for disputes to arise, it is essential to have a written contract detailing the commitments of both the beekeeper and the grower (see sample on pages 198–199).

Know the Crop

A cosmopolitan pollinator, the honey bee is known for servicing a number of crops, particularly fruits, nuts, and vegetables. A comprehensive listing is found in what many call the "pollination bible," *Insect Pollination of Cultivated Crop Plants* by S. E. McGregor, originally published by the U.S. Department of Agriculture in 1976 and no longer in print. Fortunately, this volume is currently available from a number of electronic sources. One is updated periodically (see Resources).

Each crop has its own requirements for effective pollination, and the commercial pollinator should become an expert in these and be able to communicate them to growers. The least forgiving and recently most lucrative commercial pollination opportunity is California almonds. The crop blooms so early in the season that bees have little opportunity to build populations beforehand. During

Pollination: the sweet miracle at the heart of the garden, orchard, field, and farm.

the period that the colonies are in the groves, there are precious few resources to sustain them during seed set. On top of that, the weather during the critical weeks is notoriously unreliable.

Many colonies in almonds and/or other large-scale plantings have been pushed to their nutritional limit. They may also be suffering from the ravages of mites and other organisms such as **nosema**. It is thus no coincidence, therefore, that many appear to have experienced colony collapse disorder (CCD) (see page 180).

Moving the Bees

Pollinating outfits are by necessity migratory, and operators must develop an efficient system to move bees that suits their enterprise. In small-scale operations, this can range from hauling colonies in the back seat of a car or pickup truck to using custom-designed trailers. Many of the latter have hives permanently affixed so the trailer and its colonies become a single unit that can be simply unhooked and left in or near a field. Operators of large-scale outfits generally use a combination of 18-wheeled semi trailers and forklifts.

BEEKEEPER'S STORY

I NOW HAVE ABOUT 60 HIVES, which I plan to expand to about 100 or so production hives when I retire from my day job (I own and operate a full-line pet store). I will have more time then for that many hives and they will give me, hopefully, some needed retirement income.

I sell all the honey I can produce, and have even bought some from other beekeepers to be able to get through the year. I sell about 150 pounds per month from my pet store and have two supermarkets and a health food store where I also sell a lot. I sell honey, honeycomb, chunk honey, creamed honey, beeswax, soap (purchased), and a few other purchased bee products at my pet store. I have quite a honey section.

This past spring, I pollinated almonds, cherries, pluots, and a small home orchard. I do this with a friend I helped get started in beekeeping. We move our hives by hand and with a flat trailer that holds about 30 to 40 hives. He and I put out about 100 this past spring.

After spring pollination, I move the bees to oranges. After oranges, I move them to safflower or pomegranates or wildflower areas, all within 15 miles of home (except the oranges, which are about 25 miles). Late in the season, I move to my winter yards, where cotton is grown nearby, for my last crop and fall buildup. A couple of those yards are close to town, so they get the advantage of homeowners' flowers. The wildflower areas, if not supplemented with town flowers, have been nonexistent here for the last two years. There hasn't been enough ditch irrigation water to bring up the wildflowers, so we mostly rely on town flowers and the crops I mentioned.

Laurence E. Hope, California

Managed honey bee colonies are far more valuable for pollinating plants than for supplying honey and other bee products.

A BIG BUSINESS WITH BIG CHALLENGES

According to a white paper on the health of U.S. beekeeping by the American Association of Professional Apiculturists (AAPA): "Honey bees provide essential pollination services to U.S. fruit, vegetable and seed growers, adding $8–14 billion annually to farm income and ensuring a continuous supply of healthy and affordable foods for the consumer. About 2 million colonies are rented by growers each year to service over 90 crops. The almond crop alone requires 1.3 million colonies and is predicted to require 2.12 million by 2012 (about 95 percent of all colonies currently in the United States).

"Increasing demand comes at a time when beekeepers are confronting the most serious challenges the industry has ever faced. A steady supply of healthy colonies cannot be guaranteed as parasitic mites and the rigors of migratory beekeeping continue to cause significant die-offs. A weakened beekeeping industry affects not only beekeepers, but also growers and consumers who pay higher prices for fewer goods. A weakened industry also contributes to social stress in rural America and increases our dependence on foreign sources of food. To ensure an adequate and sustainable supply of healthy honey bees, it is imperative that a new degree of cooperation be attained among researchers, beekeepers and growers, supported by elected and appointed government representatives."

Stay Local to Start

The beginning beekeeper will find it a formidable challenge to keep colonies in good shape for any period of time, let alone to ensure that they are ready to move quickly into fields for pollination. Because of the stress to colonies and other factors, the would-be small-scale commercial pollinator should concentrate on servicing local fruits and vegetables in the region. The advice given to beginning honey producers to start small and grow with experience goes in spades for anyone offering pollination services.

Keep in mind that it is extremely rare to find a crop that produces a nectar surplus while being pollinated. Beekeepers, therefore, usually must give up one goal to achieve the other.

Small-Scale Pollination

With the problems besetting beekeepers today, such as the spread of mites and concerns about colony collapse disorder, there are increasing restrictions on the interstate movement of bees. This threatens the continued availability of large numbers of colonies from the large migratory operators. Pollination services provided by small, local beekeepers are thus becoming of more interest to both beekeepers and growers. Perhaps the biggest drawback to local pollination is that most beekeepers are part-timers who individually do not have enough colonies to satisfy the demands of growers. A second problem is the quality of available colonies.

It is quite possible for several small-scale beekeepers, each with only a few colonies, to group their holdings to satisfy the needs of a particular grower. The logistics of such an endeavor may be difficult, but they can be managed. The larger problem is in fulfilling the commitment with the requisite number of hives of proper quality and configuration. Pollination contracts are usually arranged several months in advance. No grower wants to be out looking for a beekeeper a couple of weeks before bloom. Any beekeepers, therefore, who are a party to a pollination agreement must be competent enough to bring through winter the number of hives committed, and the colonies must be of adequate strength to do the job at hand.

This brings us to the heart of the matter. All colonies are not suitable for pollination. There is a preferred configuration of most commercial colonies (**units**) rented for pollination. The size and adequacy of a pollinating unit is usually expressed in terms of bees, brood, food reserves, and the condition of the hive (**queenright**) itself. Taken in that order, a colony that is suitable for commercial pollination should generally have the following attributes:

It will have a laying queen and a population of more than 30,000 bees. A lesser population requires that more of its adult bees tend brood. It thus cannot send out a sufficient field force to do the pollinating job at hand. As the population of the colony increases, so does the percentage of foragers.

It will have at least six frames with brood. The amount of brood is a reflection of the size of the adult population. A normal, healthy colony will have this amount of brood in late spring if its adult population is at least 30,000. Note that this is six frames of brood, not six frames full of brood. This is in keeping with the spherical shape of the brood nest, where the outer frames are less than full.

It will be housed in two deep bodies (or equivalent space). A colony of 30,000 bees or more requires this amount of space for normal development. Anything less may inhibit normal growth of the colony, which may in turn reduce the amount of foraging and consequently the amount of pollination. A smaller hive may also encourage swarming and the consequent loss of half the population.

Large colonies can also create problems. A colony in two hive bodies but with a large population, say 50,000 and up, may also swarm, and a hive with more than two bodies is difficult to lift and move.

It will have honey reserves in proportion to the population. Brood rearing and colony expansion depend on a certain level of food reserves present in the hive — three frames of honey, at minimum. Anything less may result in a reduction in brood rearing. A good honey reserve also allows the colony to expend a proportionally larger amount of its effort on pollen foraging. Although the bees accomplish pollination while foraging for either nectar or pollen, they are usually more efficient pollinators when foraging for pollen.

TWO WEAK OR ONE STRONG?

Some beekeepers believe that if the available colonies are not strong enough, add more. Two weak colonies, however, do not equal one strong colony. It has been well established that two colonies of 15,000 bees each are not the equivalent of one with 30,000. On a percentage basis, smaller colonies put out weaker field forces. With bee colonies, you cannot make up for poor quality with additional quantity.

HIVE PLACEMENT

Place hives away from residences and livestock, but as near as possible to the crops being pollinated.

Is Pollination for You?

Contracting hives to a grower for pollination, whether written or verbal, is a serious commitment. It is not something to be entered into lightly and unless done on a commercial scale, it is likely to conflict with what might be your primary reason for keeping bees: honey production. Crops requiring pollination often bloom from May to June, just when colonies are building up to maximum levels. In many areas, honey production is at a peak. This is not a time to put stress on hives.

Pollination requires that the hives be moved twice during this period, and more often if more than one crop is involved. Each move is a setback to colony development and production. Furthermore, hives used for pollination are often not supered for reasons of weight and ease of transportation. Storage space in the hive is, therefore, limited and the possible congestion that results may encourage swarming. Surplus honey from a hive used for pollination becomes a chancy thing.

Pollination is hard work. Orchards, cranberry bogs, and blueberry patches are often difficult to access, especially at night when most of the moving is done. And even without supers, hives are heavy and awkward. Your friends and your spouse always seem to have pressing business elsewhere at the time the hives need to be moved.

Should you decide to rent out your hives for pollination, start on a small scale and work into it slowly. It is not for the faint hearted.

KEYS TO A SUCCESSFUL COMMERCIAL POLLINATION ENTERPRISE

- Maintain strong colonies so an optimal number of foragers is continuously available.
- Know intimately the requirements of each crop to be pollinated.
- Develop a system for moving colonies quickly and efficiently.
- Insist on a signed contract.
- Recognize that honey production and commercial pollination may be mutually exclusive.

10

DISEASES AND PESTS OF THE HONEY BEE

A member of the order Hymenoptera, the honey bee is a **holometabolous** insect, which means it undergoes complete metamorphosis, its form changing drastically from egg to larva to pupa to adult throughout its development.

Besides being different in form, these life stages also live in distinct environments. The larvae eat while the pupae no longer feed but "rest" in brood cells. Adults emerge from cells and feed the larvae and forage for pollen, nectar, water, and so on. Individuals in each life stage, therefore, are generally vulnerable to specific diseases and/or pests.

Diseases and pests of the brood (bacterial, fungal, and viral maladies of the larva and pupa) do not affect the adult. Likewise, adult conditions such as nosema, viruses, amoebas, flagellates, spiroplasms, gregarines, and tracheal mites affect only that life stage. A major exception, however, is the varroa mite, which attacks both brood and adults.

Beekeeper manipulations can stress colonies and make the effects of many diseases and pests worse. Some diseases may clear up on their own since honey bee colonies have formidable defenses that they can call on (see below).

Noninfectious disorders also may occur in adult bees, and along with other diseases and pests, may result in what is called disappearing disease. Colony collapse disorder (CCD) is a recent, perhaps related phenomenon (see page 180).

Innate Defense Mechanisms

In an article in *American Bee Journal*, Jost Dustmann discussed the natural defense mechanisms protecting the health of the honey bee. He asked provocative questions: When is a social organism like the honey bee sick? Does simply the presence of pathogens inside a colony, single bees, food, or wax mean a colony is ill? To address the second question first: Probably not. Only when the number of diseased or dead bees, larvae, or pupae exceeds a certain limit and disturbs the normal functioning of the colony can these terms be applied. Indeed a small loss of infected individuals in all developmental stages must be considered normal.

The worker bee at center has one varroa mite behind its eyes and another on its back between its wings.

The honey bee as a social organism indeed has formidable defenses against pests and predators. According to Dustmann, they include:

Adult cleaning, grooming, or "hygienic" behavior. This trait is the basis for several breeding programs in the United States. Bees employ this strategy to "clean" the hive of dangerous organisms as much as possible. The behavior has two purposes.

- Single infected bees die quickly, removing themselves as a source of infection.
- Worker bees quickly identify abnormal (diseased) individuals (adults, larvae, and pupae) and eliminate them from the hive.

Hygienic behavior is shown to be effective against American foulbrood, nosema, chalkbrood, and sacbrood, and it is also responsible for removing insects who suffer from paralysis caused by viruses. The article also predicted the trait would be involved in resistance to varroa mites in European bees as in Asian species, which has turned out to be the case.

Quick regeneration after losses of population. This is essential as removal of diseased individuals can be compensated for in a short time. The ability to produce, raise, and, therefore, replace bees rapidly often generally outstrips even the greatest current threat to populations, the varroa mite.

Continuous rotation of bee generations. The rotation of brood followed by adults followed by brood is analogous to crop rotation in commercial agriculture, where growing of one crop variety is often followed by sowing of another. Pests that feed on one particular plant crop, therefore, cannot further build their populations when faced with something different the next planting season.

Swarming and absconding. A major defense is building new wax comb after bees swarm from their old colony. New comb is free from disease organisms and potential wax contamination. This appears to be a predominant strategy of Africanized honey bees, which not only swarm but also may abandon (abscond from) their nests, leaving pests, diseases, and contaminated wax comb behind. Beekeepers can mirror this activity by renovating combs periodically.

Restricting disease to either larvae or adults. This strategy, explained above, ensures that if larvae are diseased, adults are not, and vice versa. Both stone brood and varroa are exceptions to this rule.

Immune reactions. Researchers have observed the formation of bactericidal molecules (peptides) in individual bees and consumption of foreign materials (phagocytosis) by blood (haemolymph) cells. Observed in single adult bees, these reactions have not been seen across the colony.

THE PROMISE OF HYGIENIC BEHAVIOR

Hygienic bees detect diseased or mite-infested brood, quickly eliminating them from the nest. The late Dr. Walter Rothenbuhler, retired from Ohio State University, is credited with first identifying this trait and studying its effect on the incidence of American foulbrood. It is based on two genes, which he found to be heritable. The knowledge lay dormant for a number of years until being resurrected by Dr. Marla Spivak at the University of Minnesota's St. Paul campus.

BEEKEEPER'S STORY

THE EMERGENCE of the small hive beetle is the biggest change since I began. I've not had the thrill of a sliming yet, but I'm sure it's coming.

In the beginning, I tried to use all of the chemical products to control varroa, and the tracheal mite in particular, but I noticed that my queens were dying off and hives did not seem able to replace them, so I quit using any chemicals. I switched fairly early on to screened bottom boards and use an occasional sugar dusting for varroa, but that's about the extent of my integrated pest management approach.

Don't believe everything you read in the beekeeping press regarding control of small hive beetle, varroa, and other pests. The magazines will publish anything any beekeeper offers as gospel. Keep a healthy skepticism about new methods proposed, as most lack any scientific research. The good news is that beekeepers will try anything, and some of it works. Ask around.

Mark Beardsley, Georgia

Stinging and biting. The first is self-evident. Biting is a proven defense mechanism against varroa mites, but found to occur in European bees at lower levels than in Africanized and Asian honey bees.

Proventriculus and peritrophic membrane. The former structure prevents entry of foreign organisms, such as bacteria and pathogens, from the crop into the digestive system. The latter lines the digestive tract, protecting it from rough materials and also preventing entry of bacteria and fungi into the haemolymph.

Antibiotic substances. Extremely important in honey bee defenses, a wide range of these chemicals may be found in honey, stored pollen, and propolis. They are also the basis for many human health treatments, including the use of honey as a surgical dressing and consumption of propolis as preventative medicine.

The *American Bee Journal* article concluded with another question: How can the beekeeper manage bees without interfering with the above natural strategies? The author had several recommendations.

- **Select and breed bees that have the necessary defense mechanisms already in place.** Controlled mating is essential in this process.
- **Ensure that the environment provides enough of the right food for bees.** If not, then the beekeeper must feed both carbohydrate (sugar) and protein (pollen substitute/supplement).
- **Determine the beehive is the correct size with reference to colony size and management practices.** This may mean establishing artificial swarms that build new combs and uniting young colonies with older ones established the year before; such practices ensure rotation of bees and wax. Routinely replacing old combs is something every beekeeper should consider.
- **Keep bees without using chemicals.** In general, Dustmann concluded, chemical treatment of a colony will often interfere with its natural defense mechanisms. This is especially true for many antibiotics, which cannot eradicate infections but only mask symptoms, leading to reservoirs of disease ready to break out at any moment.

A CASE FOR DIAGNOSTICS

The time has come to put more emphasis on diagnostics in beekeeping. This concept is well known in many other areas. Veterinarians, state Extension Services, and private consultants routinely diagnose domestic animals and ornamental and agronomic crop plants for a wide range of ailments.

In the past, most diagnosis of bee colonies has been left to the bee inspector and beekeeper. Generally, these efforts amounted to looking for gross symptoms of brood disease. Introduction of tracheal mites and now varroa, however, has changed things. Fortunately, varroa can be seen with the naked eye, although it is helpful to have a magnifying glass or dissecting microscope for verification. Tracheal mites are another matter. The only way to find out how infested bees are is to dissect the anterior (nearest the head) tracheal branches of the thorax (the part of the bee to which are attached the wings and legs). This requires a binocular dissecting microscope.

Perhaps the most efficient way to find out what's affecting bee colonies would be to use a professional diagnostic service. Unfortunately, very few of these enterprises exist. In addition, not many guidelines help a beekeeper make wise decisions concerning what services to contract. Historically, beekeepers have been a self-reliant bunch because they could do their own disease diagnosis and control. Using a qualified diagnostic laboratory to determine the parasite and disease status of bee colonies would leave the beekeeper time to perform management necessary to get the most out of a beekeeping operation.

Unfortunately, use of a registered chemical to control varroa may be unavoidable as part of a control program in some North American apicultural situations. It is becoming clear, however, that chemical treatment for varroa may create more problems than it solves. See page 175 for further discussion about using chemicals to treat varroa.

Brood Diseases

Honey bee brood diseases are usually categorized by the kind of organism that causes them, such as bacteria, fungi, and viruses. Typically, bacteria cause the most feared diseases, American and European foulbroods. At certain times, however, the fungal disease chalkbrood can be problematic, as well as the viral-induced **sacbrood**. Mixed infections of viral and bacterial diseases are also possible.

Finally, the varroa mite and tracheal mite have been implicated in a relatively new condition known as **honey bee parasitic mite syndrome**.

Bacterial Diseases: American Foulbrood

One of the most significant honey bee diseases, **American foulbrood (AFB)**, has at times caused horrific outbreaks in much of the country. It is the prime reason bee inspection services exist in many states. Unfortunately, rules vary widely among regulatory agencies, sometimes causing more harm than good when it comes to controlling this disease.

The cause is the spore-forming bacterium, *Paenibacillus larvae* subsp. larvae (originally named *Bacillus larvae*). The spores are resistant to heat and drought and can remain dormant for many years, germinating at optimal temperature and moisture levels. AFB differs in several respects from European foulbrood (EFB) (see pages 164 and 187).

Symptoms. Typically, AFB attacks older larvae and pupae; perforated cappings are a telltale sign of the disease. It is thought that worker bees begin to remove the cappings from diseased pupae, but then abandon the dead individuals they expose. Certain colonies do a better job removing diseased pupae and are, therefore, more "hygienic" than others.

A time-honored procedure called the ropy test involves inserting a small twig into a suspected cell, mixing the contents, and then observing if the resultant material sticks together and "ropes out" when it is withdrawn. Scales, the remnants of dead pupae that have been dehydrated, are often seen adhering to the sides of cells; they are extremely difficult to remove. Affected brood also has a specific odor, sometimes characterized as a "gluepot."

Treatment. Beehive inspection and burning infected hives (wooden ware and combs of brood and honey), in combination, were the only ways to fight American foulbrood until antibiotics came along in the 1940s. Terramycin, a tetracycline product, is labeled and used extensively in the United States to control the disease. Most use is preventative, but there is evidence that the causal bacterium can and has become resistant to antibiotics, probably through long-term exposure. As of

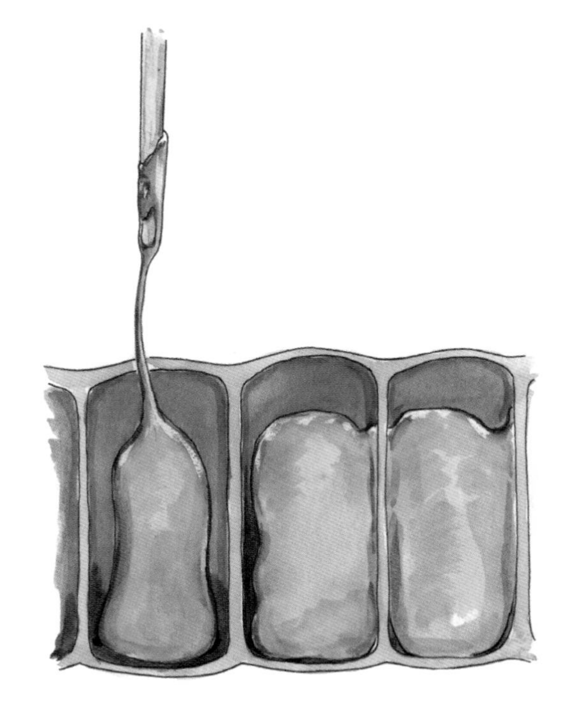

Symptoms of American foulbrood include perforated brood cappings (left) and pupal cell contents that "rope out," adhering to a stick (right), and have a distinctive caramel color.

January 1, 2017, therefore, the rules governing the use of this material have changed. No longer is it available over the counter to beekeepers. A relationship must be developed with a registered veterinarian in order to obtain this antibiotic and any others that might become available in the future.

Controlled breeding is of some value in producing hygienic bees but has not been developed to its full potential. Breeders may revisit this, however, as more begin to employ varroa-sensitive hygienic (VSH) stock. In the future, radiation treatment, essential oil application, and biological control may become treatments for AFB.

Bacterial Diseases: European Foulbrood

Distinct from American foulbrood, **European foulbrood** (EFB) is a bacterial disease of the brood caused by *Melissococcus plutonius* (formerly *Streptococcus pluton*). The disease is generally associated with stressful conditions including queenlessness, attacks by other pests, and intrusive management by the beekeeper. It is called European foulbrood because much of the work on this disease was done on the European continent. In contrast, most of the work on American foulbrood was done in the Americas. European foulbrood is almost never found in a colony that has American foulbrood.

A CASE FOR NEW COMB

Many beekeepers are thinking there might be something to the notion of periodically replacing old comb. The idea is prevalent in Europe but raises a few eyebrows in the United States. Conventional wisdom dictates that even very old comb is still serviceable. In fact, over a period of years, comb gets so strong it can be handled very roughly with little chance of destroying its integrity. So it might be argued that old comb's value increases with age.

There are advantages to replacing old comb. The cast skins of generations of bees become incorporated into the cell walls, causing the cells to shrink, and over time may result in adult bees as much as 17 percent smaller than normal. Honey also becomes discolored in dark comb. Finally, old combs can be considered a sink that harbors disease organisms, stimulates fungal growth, and is contaminated with mite-control substances.

A case might also be made for using foundation and newly drawn comb to discourage wax moth. Its larvae don't do well on foundation or new white wax, but grow fat and juicy on comb that has been darkened by an accumulation of cast larval skins, pollen, and other moth nutrients.

Comb is also easily contaminated by pesticides, which often have a chemical structure similar to that of beeswax. Now that many beekeepers use chemicals inside the beehive to control varroa mites, the situation is more critical. It is becoming clear that honey bee brood may be much more affected by sublethal doses of pesticides found in wax than previously thought.

Comb renovation is a new concept. Some researchers and beekeepers believe comb should be renovated every 3 years, but little research has taken place. It makes good sense that reworking and rotating comb should become just as much a part of any beekeeper's management as preventing swarming or preparing colonies for winter.

Symptoms. The symptoms of EFB are often not definitive and the disease is considered less problematic than American foulbrood. Often the only way to determine whether the disease is present is by microscopic examination.

In general, EFB attacks younger larvae, so perforated cappings are not seen. It is easier to remove the resultant scales than those of AFB.

Treatment. Requeening can often clear up this disease by providing a break in the brood cycle, giving the bees a chance to clean out infested cells. There is evidence that the chalkbrood fungus (see below) inhibits growth of European foulbrood, and comb renovation may be an important technique to combat this disease. As with AFB, it is possible to treat EFB with the antibiotic Terramycin.

Fungal Diseases: Chalkbrood

Caused by the fungus *Ascosphaera apis*, **chalkbrood** is sporadic, thought to be stress related, and often found in low levels within colonies. Generally, the disease occurs in spring in European honey bee colonies. It can also be a problem among Africanized honey bees in the American tropics, particularly in drone brood at high altitudes. The disease may occur in association with other fungal diseases.

Symptoms. Affected larvae resemble pieces of chalk, hence the name. Ranging in color from white to green and gray, these "mummies" often appear near the entrance to colonies because the workers have removed them. Chalkbrood produces spores that persist for long periods and give the mummies their characteristic colors. When fed to healthy colonies, pollen contaminated with chalkbrood mummies incites the disease.

Treatment. There is no known treatment for chalkbrood. Because it is a fungus, keeping colonies dry through proper ventilation is a good preventative measure. Comb replacement can also reduce the potential numbers of microorganisms in a colony. Requeening with hygienic bees may help. Finally, stress on bee colonies is known to be responsible for chalkbrood outbreaks.

Chalkbrood produces "mummies" of different colors, which are often ejected from the colony.

Viral Diseases: Sacbrood

Sacbrood, a viral disease found sporadically in honey bee colonies, is not considered a major problem. It may occur in conjunction with other brood diseases.

Symptoms. Symptoms may appear similar to those of American foulbrood, including perforated cappings. The "ropy test" (see page 163) can often be used to discriminate sacbrood from AFB. Unlike the scale of AFB-affected brood, larvae with sacbrood appear like a sac of fluid, hence the name.

Treatment. There is no known prevention or treatment.

POISONING

Plant and pesticide poisoning occur in both honey bee brood and adults.

PLANT POISONING

Plant poisoning of brood and adult bees is a sporadic problem that occurs when bees collect either the pollen or nectar of certain plants that are toxic to them. This can be a special problem throughout the Americas when plants that are not compatible with introduced honey bees bloom in profusion. Potential problem plants include:

- Summer titi (*Cyrilla racemiflora*), which can result in a phenomenon called "purple brood" in the southeastern United States.
- Carolina jessamine (*Gelsemium sempervirens*), also found in the Southeast.
- Some milkweed plants (*Asclepias* spp.)

PESTICIDE POISONING

If foragers bring agricultural pesticide residue back to the hive, the nurse bees may die after feeding on contaminated honey or pollen, and the brood will exhibit symptoms of neglect. The symptoms of poisoned honey bees often depend on the class of pesticide involved.

Beekeepers in areas where pesticides are applied to various crops should be in close contact with farmers who are applying these chemicals.

Increasingly problematic is the use of pesticides by beekeepers themselves as they attempt to control parasitic mites in colonies. More and more evidence shows that interaction and synergy among chemicals in the hive and in the field are growing problems as more and more of these materials become available and used.

Symptoms. The clearest indication of serious pesticide poisoning is the existence of a pile of dead adult bees at a colony's entrance. In many instances, however, this may not be a definitive symptom. Many poisoned bees are often lost in the field before returning to the colony.

Pesticide poisoning can also result in a loss or absence of bees within a colony, sometimes characterized as "disappearing disease" and, more recently, colony collapse disorder (see page 180). Cooperative Extension Service offices have published several guides on protecting honey bees from pesticides, available on the Internet and in print (see Resources).

Adult Diseases

As those of the brood, adult diseases in honey bees generally affect only that life stage. The principal illnesses are caused by nosema and several viruses associated with parasitic mites. A few other organisms, such as amoebas, flagellates, spiroplasms, and gregarines, have been found to affect adults, but rarely are considered a major problem. Again, as in brood diseases, many of these conditions are also made worse by stress put on colonies through beekeeper manipulation.

Nosema

Nosema apis is a spore-forming microsporidian parasite of the adult honey bee's digestive system. It is not strictly a disease, although in Europe and elsewhere it may be called nosemosis; rather, it is a condition that occurs in honey bees when the organism reaches epidemic proportions.

Nosema attacks the cells of the gut lining, leading to many problems, including reduced absorption of nutrients and dysentery. The organism may also activate viruses, resulting in symptoms sometimes associated with tracheal mites (page 169).

Symptoms. Many beekeepers call nosema the "silent killer." The disease is often discounted because it doesn't appear to kill colonies outright. Diagnosis is not a simple process because the symptoms are not always definitive. They can include K-wings (page 170) and dysentery, but not always.

Treatment. The one material known to control nosema effectively, fumagillin (sold under the brand name Fumagilin-B), is expensive, and recent information suggests that this antibiotic treatment may be harder on bees than originally thought. Thus, using fumagillin carries a risk and should be assessed accordingly. Acetic acid has been used to destroy the viability of nosema spores on stored comb.

BEEKEEPER'S STORY

SOME KEEPERS prefer Italian bees. My preference is for the darker side, including Russian or Carnica breeds and the "survivor" critters of local origin. Clear-cutting of old-growth forests has contributed to the demise of feral bees. There are a few left here or there, and if possible, one tries to catch and propagate swarms from those survivors.

My colonies have not been affected by colony collapse disorder or the late curse of "nosema" — at least not that I am aware of. Neither did I experience losses by varroa to any remarkable degree. I tried pesticides — everything from essential oils to chemicals — all of which proved detrimental or useless.

I prefer the screened bottom board as a mite-disposal method. Whether a screened bottom board is instrumental to treat varroa is, of course, beyond my means to judge. Naturally, I saw varroa on the bottom board beneath the screen. The method is conducive to trap wax moth larvae because the wax debris falls through the screen. I have had no trouble with this insect at all.

H. E. Garz, Washington

MANAGING STRESS

Beekeepers often ignore the influence of various kinds of stress on a colony. Stress doesn't cause a particular disease but weakens a colony so that disease and parasites can get a foothold. A bee colony should be looked at as a "black box" into which energy (pollen, nectar, water) flows and is then converted into products (honey, beeswax, brood, bees). Anything reducing energy flow into the box is termed stress. The beekeeper is interested in removing as much energy from the system (honey, wax, or pollination in the form of worker bee energy) as possible. Paradoxically, therefore, the beekeeper becomes a component of stress on a colony.

Weather is perhaps the greatest source of potential colony stress. Inclement conditions slow down or stop altogether the flow of nectar and pollen into a colony. In spring and early summer, the brood consumes large amounts of energy (food). A sudden shutdown of foraging (because of rainy conditions or confinement by the beekeeper for moving, for example), if prolonged, causes severe stress. The bees' solution is to stop brood rearing, reducing energy investment and colony growth.

Good management practices that provide food, water, ventilation, and a minimum of colony disturbance result in less stress and more potential colony productivity. These are the goals of every beekeeper.

A new nosema has been detected in U.S. honey bee populations recently, although apparently it has been present since the 1980s. *Nosema ceranae*, which first affected Asian honey bees (*Apis cerana*), has a different etiology than the traditional variant, *Nosema apis*, and may be much more damaging. It has been directly implicated in colony collapse disorder in Spain.

Parasitic Mites

Several parasitic mites have been found to be damaging to honey bees, and problems are most likely to arise when they are introduced from one area of the world to another. It is difficult to determine how introduced organisms might interact with potential hosts in a new environment. This result of novel, unexpected behaviors in new environments has been called the nemesis effect. One example, as noted previously, is the wide-ranging effects of the exotic Western honey bee, introduced by Europeans to the Americas. Although now considered a part of the North American landscape, the changes wrought by *Apis mellifera*, in conjunction with the plants brought over to sustain it, resulted in a transformed landscape that was devastating at the time to American Indian aboriginals and many existing native plants.

For beekeepers in North America, introduction of both the tracheal mite in 1984 and the varroa mite in 1987 has been responsible for large-scale colony loss, and implicated in a new condition known as honey bee parasitic mite syndrome (HBPMS). Of the two, the varroa mite is the most pernicious and damaging. It appears to have jumped from its original host, the Asian honey bee, to the Western honey bee when colonies of the latter were purposefully moved into territory occupied by the former. As noted elsewhere in this volume, varroa must now be considered a permanent part of the Western honey bee colony in areas where it exists.

It is important to realize that the possible introduction of other mites remains a constant threat to worldwide beekeeping. Perhaps the best potential example at present is *Tropilaelaps clareae*, a mite currently found on the Asiatic bee, *A. dorsata*, sometimes called the "giant honey bee." Other variants of varroa may also be candidates for introduction. Honey bee stock introduction must, therefore, be carefully considered and closely regulated to achieve a diversity of genetic stock with minimal risk of spreading other diseases and pests around the globe.

Unfortunately, all of this creates confusion when clear thinking is absolutely necessary. The result is that some beekeepers have hinted at taking action themselves by illegally importing bees from Europe, Latin America, and Africa. Several extant Asiatic honey bee mites might be candidates for accidental introduction.

Perhaps the most worrisome honey bee is the cape bee from South Africa (*Apis mellifera capensis*), which could cause chaos in existing North American populations. We can't yet predict the effects of transporting other bees such as *Apis cerana* and *Apis florea* to other lands. In general, we should bring in honey bees only after a good amount of debate and critical thinking. No one believes that random introduction of stock without some safeguards is anything but a prescription for disaster.

INTRODUCING BEE STOCK

Since introduction of tracheal and varroa mites in the 1980s, there has been much discussion about importing bee stock (see pages 179–180). The USDA has imported Russian stock over the last decade, which is considered tolerant to varroa. Stock from England has also been brought in for resistance to tracheal mites. The latest introduction has been Australian stock in response to honey bee shortages for pollinating California almonds. Other candidates for introduction also exist, including nonnative species from Asia or Africa.

Many voices have added to the importation controversy over the years. Although intentions are good, the specifics for proposed introductions are not always clear. Surely we need resistant stock, but is importation the only solution? Under what conditions will it be beneficial or harmful to the industry? Do we really want to import bees with the public relations image of the African strain? Many suggest we cannot wait and advocate bringing in stock no matter the future cost; others say this would be a great disservice to the beekeeping industry.

Tracheal Mites

Three *Acarapis* mite species are associated with adult honey bees: *A. woodi*, *A. externus*, and *A. dorsalis*. They are difficult to detect and identify because of their small size and similarity. They have, therefore, frequently been identified only by their location on the bee. Only *A. woodi*, or the tracheal mite, has been found to be detrimental to honey bees. It lives in the honey bee's breathing apparatus, the tracheal system, and leaves only to migrate to other bees.

First detected in 1984 in the United States, the tracheal mite was held responsible for up to 90 percent colony loss in some regions of the country. Beekeepers so feared the mite's introduction after an outbreak in the United Kingdom of what was called "Isle of Wight" disease that they supported passage of a 1922 U.S. law restricting the importation of honey bees from anywhere else in the world. The Honey Bee Act has been amended several times over the years, but it was unable to keep out the tracheal mite.

Symptoms. Tracheal mites are far more difficult to detect than varroa. They cannot be seen without a microscope and each potentially affected bee must be dissected to conclusively determine their presence. Certain symptoms, however, suggest a high infestation of tracheal mites. They include:

- Bees crawling in front of the hive entrance, unable to fly
- Poor clustering in cold weather
- Individual bees with K-wings (when the wings and the body are shaped like the letter K)

If a colony dies from tracheal mites, bees may be found randomly throughout the hive bodies (instead of in a cluster), usually with plenty of honey on hand. K-wings, another symptom, are a phenomenon in which the wings are unable to fold correctly and instead stick out almost vertically from the body.

Treatment. Menthol in grease (vegetable oil) patties is used as a tracheal mite treatment in regions where the pests are problematic. The best time to treat for these mites is late summer and early fall when the outside temperature is warm enough to promote menthol evaporation.

Some beekeepers claim that some chemical treatment for varroa (page 175) can also control tracheal mites, but this has not been intensively studied.

Worker bee with K-wing

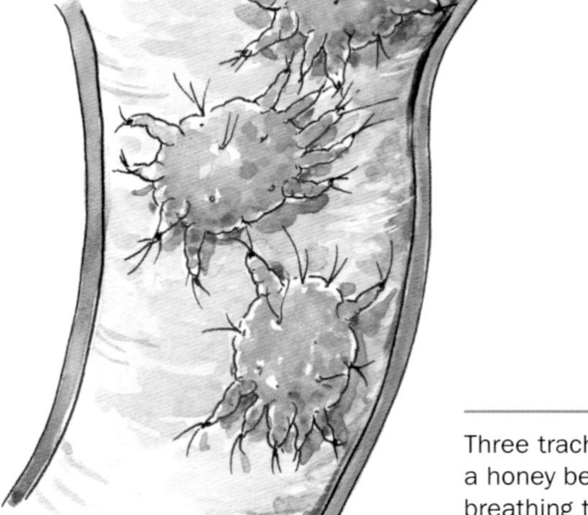

Three tracheal mites inside a honey bee's tracheal or breathing tube.

THE TRACHEAL MITE ENIGMA

There are a good many beekeepers and scientists scratching their heads about tracheal mites. Nobody seems to be able to reconcile why high mite levels in fall do not necessarily translate into large colony losses in spring. Informal reports from California to Michigan reveal this quandary. One prominent researcher summed it up: "I still have not figured out the tracheal mite. . . ."

Other bits of evidence add to the confusion. It seems that researchers in the Rio Grande Valley of Texas, where the first detection occurred in 1984, are unable to keep mite levels high enough to do effective research on the critter. Early studies in Florida were also plagued with very large variation in mite populations that confounded investigators.

Tracheal mite infestation and its correlation to colony damage is just one situation that parallels similar conditions in beekeeping and other agricultural enterprises. Many of these appear to ebb and flow over the years, often with no particular pattern. Prevailing environmental conditions will favor certain disease or pest problems in particular seasons, but not others, making predictions difficult. The human manager can only hope to stay a little ahead of the curve in determining the next biological challenge faced by honey bee colonies.

Over time, the tracheal mite has been determined to be more of a problem in temperate areas, where honey bees are confined for long periods during harsh weather, than it is in subtropical regions. It has a cyclical population, which includes a fall buildup, a winter peak, and a summer crash. Although many beekeepers consider the mites problematic, others ignore their presence, principally because of the rigors of the detection process.

Breeding for Tracheal Mite Tolerance

There is considerable evidence that many honey bees throughout the southern United States have become somewhat tolerant to the honey bee tracheal mite, and no monitoring or treatment is necessary. Breeding for tolerant bees has been an important strategy in some temperate areas. Perhaps the best example is a program that continues to this day in Ontario, Canada, where beekeepers subject honey bees to a "quick test" for tracheal mite tolerance and then select for this characteristic in a breeding program.

This test measures the ability of the colony to resist infestation from the tracheal mite, which lives in the bee's tracheal or breathing tubes. In order to perform this test, the beekeeper must remove a frame of emerging bees and send it, along with some bees, to the Tech-Transfer Team working for the Ontario Beekeepers' Association. The adhering bees are brushed off and the frame is incubated overnight in a special cage to trap all emerging bees. The bees are tagged to identify which breeder queen each came from and then introduced randomly into infected colonies.

The newly emerging bees are susceptible to tracheal mites in the first seven days. A week later, the bees are retrieved from the colonies and classified according to their original location (mother-breeder queen). Each bee is then sent to the bee lab to be sliced and examined under a microscope to count the number of tracheal mites it contains.

Varroa Mites

The varroa mite has been extensively described in this volume as one of the permanent residents in the hive. From that discussion, it is safe to conclude the following:

- Increased brood production directly results in a subsequent growth of the varroa population.
- Beekeepers can feed, combine, and split colonies, influencing both the honey bee and the varroa mite population.
- Management of the population levels of varroa, the fourth type of individual making up a colony, is as important as any other phase of beekeeping.

Recent reports of colonies lost because of deterioration of honey bee health, often classified under the rubric of "disappearing disease" or "colony collapse disorder" (CCD), have pointed to a number of causes, including viruses, nosema, pesticides, and inadequate nutrition. All of these stress factors are far worse in bee colonies with high varroa mite levels.

The message is clear: varroa mite control must come first. Without it, any and all efforts to maintain strong colonies by the beekeeper are compromised.

Management of varroa consists of two ongoing tasks: monitoring the mite population as it ebbs and flows in a honey bee colony, and taking measures to control the varroa population once it appears to be getting out of control.

Monitoring Varroa Mites

It is indeed fortunate that varroa mites can be seen with the naked eye. Estimating their number in a colony is not easy, however, and this has led to a tendency to treat even if only a single mite is seen. Several techniques are available to help the beekeeper estimate mite population. These include:

Sticky Board Detection

This technique allows passive monitoring of mite population levels. You can purchase sticky boards from bee-supply dealers or make them from adhesive-covered poster board.

1. Place the sticky board on the hive floor.
2. Optionally, you can place an 8-mesh screen (0.1 × 0.1 inch) above the sticky board to keep bees from contacting it. Mites that fall off during the bees' grooming process (or for other reasons) pass through the screen cover and adhere to the sticky board.
3. Leave the sticky board in the hive for 24 hours and then remove it to observe "natural mite fall."

Nondestructive Sampling Using Sugar Shake

This procedure will separate up to 50 percent of the mites from the insects, and it will not kill the bees.

1. Use a wide-mouth Mason jar, replacing its inner lid with an 8-mesh screen (0.1 × 0.1 inch).
2. Brush or shake approximately 100 to 200 worker bees from near the middle of the hive into the jar. Place the mesh screen on top.
3. Sift a heaping tablespoon of powdered sugar into the jar through the mesh screen. Roll the jar from side to side to

distribute the sugar over all of the bees for about a minute. Pour the sugar and dislodged mites through the screened cover onto cheesecloth.

4. Separate the mites from the sugar by sifting the mixture through a screened cloth, leaving the mites on the cloth surface.
5. Dump the bees in front of the colony.
6. Discard the mites anywhere outside the hive; they cannot survive long separated from a bee colony.

Destructive Sampling Using Ether Roll

This technique for counting mites will separate up to 90 percent of the mites from the bees, but it will kill your sample bees in the process. This is considered the best detection method at present.

1. Use a wide-mouth Mason jar.
2. Take 100 to 200 worker bees from the middle of the hive and place them in the jar.
3. Spray a short burst (about one second) of engine starter fluid (ether) into the jar and roll the jar from side to side.
4. Varroa mites will fall off the bees and adhere to the sides of the jar, where you can count them.

Varroa mites prefer drone brood. They are not uniformly present in comb, however, so counting mites in this manner is usually not effective if you wish to compare the results directly with those of other methods.

SUGAR SHAKE MITE SAMPLING

From top to bottom: Sift powdered sugar onto bees in jar; roll to distribute sugar; shake out sugar and mites together onto cloth.

An infestation level of less than 5 percent (5 infected brood out of 100 examined) indicates sufficiently low mite numbers, making treatment unwarranted. A level of 25 percent (25 out of 100 examined) or more infested brood indicates a severe infestation, which requires immediate treatment.

Sampling Using Cappings Scratcher

1. Locate drone brood by the cappings that bulge up from the cells.
2. Insert a cappings scratcher below the caps and pry them out of the comb with brood impaled.

A cappings scratcher can be used to impale pupae in drone brood. Once removed, they are examined for mites.

BEEKEEPER'S STORY

MY GRANDFATHER HAD BEES, and I remember watching him work. About 1990, I bought a farm on which a commercial beekeeper had beehives on pallets. I took the beginning beekeeper class and when he left, I started my own. The biggest thing I learned, and still believe, is how fascinating and complex a honey bee is.

Lincoln, Nebraska, has highly variable weather. This year has been wet and cool, and we are not making much honey. My hives are all two deeps, and I add four supers in mid-May. Yellow sweet clover is the primary plant my bees use. This year it was too cold. Farmers no longer allow smartweed to grow in the wheat stubble, which always filled the hives up for fall. Now, sometimes we have to feed.

Every hive has mites, so we count and treat. All the states around us have small hive beetles, so that is coming, and we are heavily infested with *Nosema ceranae*, so we are feeding Fumidil more often. All of this is very expensive, and I think chemical treatment is the reason why it is harder to produce strong hives.

Start slow. The biggest joke is the guy who decides to be a beekeeper, buys 10 to 20 hives, loses them all, and now has equipment for sale.

Dave Hamilton, Nebraska

VARROA MITE THRESHOLDS

A basic building block of integrated pest management (IPM) systems is to treat with chemicals only when the pest population exceeds some defined economic level. Below this threshold, potential damage to the crop is less costly than treatment.

Unfortunately, the economic treatment threshold for varroa mites in a honey bee colony in many areas has not fully been determined. A major problem is that detection procedures are not uniform, so interpreting the results is variable. Furthermore, a large number of materials have been approved for control, and research comparing their effectiveness, application, and other issues is underway but not complete. (See Resources for the current recommendations by The Honey Bee Health Coalition on varroa control.)

Given the above uncertainty, the regulatory threshold in many beekeepers' minds can often be a single mite in an ether roll jar or on the bottom board of one colony. This encourages overapplication of materials, leading to mite resistance, creating further uncertainty. Unless reasonable thresholds are specifically determined in localized areas, beekeepers and regulators are disadvantaged in determining any course of action to control one of beekeeping's most serious problems.

Treating Varroa Mites with Chemicals

Since introduction of the varroa mite, many beekeepers have adopted a short-term "zero mite tolerance" attitude, which has often worked to their own and the honey bees' long-term detriment. The prevailing mindset was (and continues to be in some circles) that the mite can be eradicated or eliminated from colonies. This idea has been accompanied by the use of "hard" pesticides, which include fluvalinate (Apistan), coumaphos (CheckMite), and amitraz (Apivar). These materials did indeed reduce the varroa mite level to less-than-detectible levels for a number of years, but now they are not effective in many places because of the emergence of resistant mites. Although the amitraz product has not been available since 1988, the fact that mites are resistant to this active ingredient indicates beekeepers have used it "off label" for a long period.

Any product designed to kill mites must be used according to its label. The wording on the label is the law. This book will, therefore, give no instructions on specific chemical use.

The above hard pesticide products were indeed magic bullets. They killed more than 90 percent of mites and could be employed under a wide range of environmental conditions. Their success was short-lived, however, and in many areas they are no longer effective. One reason for this is that beekeepers often didn't establish whether the level of infestation was high enough to warrant treatment. In addition, not many followed up to see how effective treatments were. This contributed to overuse of treatment products. As a result, an epidemic of comb contamination by chemicals arose in some operations. Consequently, many beekeepers are now switching to using plastic frames and foundation.

This era of the hard pesticide is now giving way to essential oils and organic materials, such as formic and oxalic acids. These "soft" chemicals often already exist in honey, reducing contamination concerns. They generally do not bioaccumulate (collect) in the comb, and the potential for development of mite resistance is thought to be reduced. On the other hand, such chemicals are generally not as effective, killing an average of less than 70 percent of mites, instead of over 90 percent like their hard cousins.

Two products made with the essential oil thymol are presently labeled Apiguard and ApiLife VAR, and one formulated with formic acid is called MiteAway Quick Strips (MAQS). These products have restricted temperature windows and require thought, planning, and care in their application.

Another treatment, Sucrocide (sucrose octanoate), uses sugar esters. In the future we can expect to see more potential chemicals, like Hivastan (fenpyroximate), so far available only in certain states with a temporary emergency use permit under Section 18 of the Federal Insecticide, Fungicide, and Rodenticide Act (FIFRA).

Again, it cannot be overemphasized: follow instructions printed on all labeled materials to the letter. **The label is the law!**

UNDERSTANDING VARROA

The number and location of mites in a colony vary based on the time of year. The varroa population is usually lowest in spring, increases during summer, and is highest in fall. During spring and summer, most mites are found on the brood (especially drone brood). In late fall and winter, many attach to adult worker bees as brood rearing decreases. Presence of mites on both brood and adult bees creates difficulty in determining a colony's exact varroa population level. The fact that some detection techniques can double as treatments also complicates calculating how many mites are found in a specific colony.

Research to develop better varroa population management techniques is continuing. Beekeepers should keep abreast of current work in this critical area as reported by their peers and researchers.

The Community Approach

When varroa became widespread, many regulators and beekeepers recommended that the best strategy to control mites, while keeping available pesticides from becoming ineffective due to potential resistance, was to implement a large-scale treatment regimen. All beekeepers treating at the same time over a wide area would preclude one of the most significant problems associated with varroa control: the reinfesting of treated colonies by mites from nearby apiaries that have not been treated. This strategy has been successful in Israel where the beekeeping community is strictly self-regulated, but it has not been seriously implemented in most of the rest of the world.

Mites and Viruses

When varroa first made its appearance, a major consideration was the mite's parasitic nature. Clearly, brood that had been parasitized in the cell before emergence could not be as productive as adults that developed from untouched larvae and pupae. One early indication of affected adults was abnormal-looking,

CONTAMINATED BEESWAX: FOULING THE COLLECTIVE NEST

Most beekeeping products are consumable items. They can be looked at as transitory (i.e., as short-term consumables on the beekeeping balance sheet). One, however, has traditionally been in the long-term-asset category. This is beeswax, that marvelous substance only the honey bee can produce. Although it can be converted into other products (candles, cosmetics), a huge amount is recycled by the beekeeping industry and given back to the industrious insects that made it in the first place as foundation.

Ever since beekeepers began using pesticides inside living bee colonies for mite control, there have been concerns about contamination. Most had to do with honey. Few considered the possible effect of long-term, widespread use of pesticides on the beeswax supply. High residue levels of pesticides, however, have and continue to be found in combs, so that pesticide use is now coming under more scrutiny. The contaminating molecules bond with the wax, making them almost unremovable during recycling.

There is some hope that beeswax from places such as Africa that don't have active varroa treatment would dilute the worldwide contaminated beeswax supply over time. In the meantime, many beekeepers have begun substituting plastic for natural beeswax foundation.

shriveled wings. Another was the difference in size between parasitized and normal adults.

More recently, however, concern has been raised about the varroa mite's relationship with honey bee viruses. There is increasing evidence that the mite's presence has resulted in significant change in the type and prevalence of viruses causing honey bee colony mortality. This is primarily because varroa has provided new routes of transmission for naturally occurring, endemic virus infections by feeding directly on brood and adults. Female mites pierce the thin membranous areas of the adult bee's body or pupal skin to feed on the fat bodies or perhaps blood. At the feeding site, there is potentially some exchange of fluids between the parasite and its host, providing a convenient entry point for viruses.

Although the varroa mite is implicated in the spread of viruses because its mouthparts effectively act like a syringe, it is not the only organism that can do this job. So can the tracheal mite, which makes holes in the integument (skin) lining the tracheae (breathing tubes) when feeding on bee blood.

The realization that viruses are becoming epidemic has stimulated a more intense examination of the honey bee's immune strategy. It seems reasonable to suggest that vaccines or genetic engineering efforts could be used to help bees develop stronger immune systems to ward off current and future viral threats.

In conclusion, no longer can the varroa mite be considered simply a parasite, and its damage assessed solely by numbers of individuals directly affected. Its connection with the spread of viruses means that beekeepers and researchers must continually reconsider their approach to varroa control.

HOW VIRUSES TRAVEL

Mites don't necessarily have to introduce viruses. Many, it appears, are generally present in honey bee populations, but are benign. These "latent" viruses appear to be activated by mite feeding, which pierces the bee's skin. Honey bee viruses can also be transmitted in two major directions, horizontally and vertically. The former is the case with most contagious human diseases where susceptible individuals are in contact with each other, such as with respiratory flu viruses (spread by sneezing) or HIV (exchange of body fluids). Vertical dispersion is intergenerational, such as between the mother and her egg. Honey bee viruses have been documented to spread from bee to bee, bee to mite, mite to mite, and mite to brood.

Integrated Pest Management (IPM)

Beekeepers are finally beginning to take a page out of other agriculturalists' books and coming around to a philosophy of IPM for varroa mite control. This technology seeks to minimize pesticide use and relies on treating only when levels of mites reach a certain damaging level or threshold. Unfortunately, the treatment threshold level is variable and in many areas it is not determined with certainty. This encourages a treatment philosophy of "just in case" application, when it may not be needed. Other parts of the IPM technology include trapping, biomechanical techniques, and brood nest management, as well as chemicals. Some combination of any or all is possible.

Biomechanical Control

Beekeepers use several biomechanical control methods as a substitute for chemical treatment.

Trapping

Many small-scale beekeepers trap varroa in drone brood, which the mites prefer to infest. To do this, insert a comb of drone cells or drone-sized foundation into a nest, allowing mites to enter the resulting brood. Then remove and freeze or kill the capped brood. Over time this will reduce the general mite population. As investigators continue to work on varroa-trapping methods in the honey bee nest, there might be further breakthroughs in this area in the near future.

Open or Screened Bottom Board

The preferred biomechanical technique for varroa control at the moment is the open or screened bottom board. Mites frequently drop off bees, due to worker grooming activities and/or the beekeeper's smoking and manipulating colonies. With a solid bottom board, the mites are not eliminated from the hive. They soon find their way back to reinfest the colony. With an open or screened bottom board, in contrast, mites that are groomed off or drop from bees fall directly onto the ground and cannot return to the hive.

You can employ this device in combination with sugar dusting, described in the section on detecting mites (pages 172–173), or in conjunction with hard or soft pesticides.

Brood Nest Management

Active brood nest management by the beekeeper can also be an effective varroa control tactic. Any break in the honey bee brood

The screened bottom board eliminates mites from a colony by allowing them to fall to the ground, where they cannot return to the hive.

cycle will also disrupt mite population growth. Creative beekeepers can try requeening, caging queens, splitting full colonies into nuclei, and other techniques that interrupt the brood cycle.

The most sustainable long-term solution is to manage the varroa mite population along with the bees. The best approach is a combination of tolerant or resistant stock, coupled with drone traps, screened bottom boards, and/or active brood nest management. Chemicals are less and less sustainable within this mix. The only way to exit the pesticide treadmill is to get off. But that's more easily said than done in the current beekeeping environment.

The Global Perspective

Beekeepers in South Africa, Cuba, and Brazil have few varroa mite problems and do not treat honey bee colonies with pesticides. These countries made a collective decision not to treat at the outset when varroa was detected. Mite tolerance was established quickly in a very few years. In contrast, U.S. beekeepers continue to have serious varroa problems, a legacy of two decades of futile efforts to eliminate mites through chemical treatment.

Tolerant or Resistant Stock

Fortunately, beekeepers in the United States now have access to an increasing variety of varroa-resistant stock. Several breeding programs have this as their goal, such as those for Minnesota hygienics and Russians. Other stocks are actively being developed based on untreated survivor stock. Indeed, one investigator, Wyatt A. Mangum, recently reported about his efforts in selecting for varroa resistance: "Clearly this stock offers a substantial and real hope for beekeepers to begin to regard varroa as a minor and ignorable inhabitant of the hive."

Again, it is important to remember that varroa introduction was only a few short years ago when compared with the long history of beekeeping and honey bee evolution. Both the honey bee and the beekeeper are still reeling from the appearance of this nonnative organism for which neither has an innate defensive strategy.

The IPM techniques listed here, along with treatment protocols and products, are continuing to evolve, and many questions persist about the effectiveness of these tools as part of any specific management protocol. A feeling of uncertainty has always been a part of the beekeeping experience, but the fact remains that it is far more risky to be a honey bee or a beekeeper in this modern, varroa-infested environment.

Russian Bees: Survivors Tolerant of Varroa

Bees from Russia's far eastern Primorsky region, which have never been treated and yet have survived decades of varroa infestation, were introduced into the United States by the U.S. Department of Agriculture several years ago and are now being propagated. The resultant Russian Honey Bee Breeders Association is pursuing a certification program and has a growing membership. It does not claim the bees are "varroa proof," only tolerant to some degree, which means they often require less chemical treatment. This advantage, however, can be voided if Russian bees are kept near other bees that are heavily infested with mites.

Changing to Russian bees has been difficult for some because these insects have a different suite of behaviors that require adjustments by the beekeeper. They are more responsive to environmental conditions than Italian bees, often being compared to Carniolan bees. Early in spring, the brood nest remains somewhat smaller than with Italian bees. But once brood rearing begins in earnest, the population can often surpass that of Italians. This quick population expansion, however, requires a more judicious approach to supering, or swarming will result. Russian bees are not for everyone, the association says, but are a good alternative for those interested in doing something a little different.

VARROA-SENSITIVE HYGIENE: ONE KEY TO VARROA RESISTANCE

In the 1990s, researchers discovered that in some honey bee colonies, varroa was not reproducing as much, providing a general degree of resistance or tolerance to the mites. At the time, this trait was named suppressed mite reproduction (SMR). This has now been renamed varroa-sensitive hygiene (VSH). Bees having the trait remove mites that have begun to reproduce, but leave nonreproductive (sterile) individuals alone.

Populations of VSH bees appear to exist around the country, and breeders and others have attempted to find and preserve the trait. Expect to see various kinds of "hygienic" or VSH bees offered to the general public in the future. Perhaps best known at the moment is a line developed at the University of Minnesota. These "Minnesota Hygienics" are being propagated by several breeders, and a program to certify this stock is in the works.

Colony Collapse Disorder

The newest phenomenon in beekeeping circles is **colony collapse disorder (CCD)**. First identified in 2006, it was responsible for the death of many hives, especially those managed by large-scale commercial pollinators. The resultant publicity often gave dire predictions — and rightly so — of what might happen to the food supply should honey bees not be available for pollination. It created enough awareness that several large-scale honey bee research programs were funded in universities and at the U.S. Department of Agriculture.

ON EXPERIMENTING

Beekeepers have always been experimenters. In many countries of the world, they are all too often experimenting with a wide range of chemicals to control outbreaks of diseases or pests. The reasons that bee researchers and regulators blanch at the thought of this activity range from a wish to obtain infested, undisturbed colonies for study to fear for beekeepers' and honey crops' safety.

Many beekeepers have been understandably frustrated by the lack of registered materials to control a number of diseases and pests. Behind any adequate study, however, lies the real world of making the many decisions needed to determine effectiveness of any chemical treatment on a complex insect society. This process is costly and time consuming because a great number of questions can arise for which there are few answers.

No experiment is worth much without an untreated control, used to compare and determine the effectiveness of a treatment. For honey bees, this should be a colony in the same state genetically and quantitatively (equal stores, amount of brood), and infested to the same degree as the one being treated. Developing adequate controls is often the most difficult part of any honey bee experiment. Finally, laboratory and small-scale experiments must be carried to the field and repeatedly conducted before being recommended on a larger scale.

Does this mean that experimentation is something best left to experts? Not necessarily. The small-scale operator has all the material needed to effectively and safely tinker with the honey bee's genetic makeup. Selecting colonies that survive apparent disease or pest outbreaks and are superior producers, and then propagating queens from them, is the long-range experimentation strategy properly in the realm of the beekeeper.

The CCD name was created to differentiate the phenomenon from a similar one called "disappearing disease," or DD. The symptoms of CCD appear to be more specific than those of DD, including:

- **Complete absence of adult bees in colonies,** with little or no buildup of dead bees in or around the colonies.
- **Presence of capped brood in colonies.** Bees normally will not abandon a hive until the capped brood has emerged.
- **Presence of food stores, both honey and bee pollen,** that are not immediately robbed by other bees and may also be ignored by wax moth and small hive beetle alike.

Precursor symptoms that may arise before the final colony collapse include:

- A workforce insufficient to maintain the brood that is present
- A workforce that seems to consist only of young adult bees
- Colony members who are reluctant to consume provided feed, such as sugar syrup and protein supplement

Since its appearance, no one single cause for CCD has been identified. Instead, most now see it as the possible result of a number of things, perhaps acting together. These include pesticides (used by growers and beekeepers), viruses, and a new nosema variant (*N. ceranae*), all of which may be additive and exacerbated by varroa. Another way to look at this is to examine how much honey bees can be actively managed by beekeepers in search of maximum production before the colonies finally collapse. This line of inquiry encourages beekeepers and researchers to look at the honey bee colony in a more comprehensive way under the rubric of "bee health."

The CCD phenomenon is not uniformly present and appears to have diminished greatly since 2010. Like many historical colony losses, it has been largely unexplained. Large-scale commercial pollinators first reported it, and it appears to be of less concern to many stay-at-home, small-scale beekeepers. Because CCD is a new phenomenon, beekeepers must keep up with current research on the topic to determine how it might affect them. One of the side effects of the media uproar over CCD has been an increased interest by the general public in taking up the beekeeping craft.

At this writing, the CCD situation continues to look more and more like an elaborate jigsaw puzzle. Unfortunately, just as part of it begins to take shape, more pieces emerge to complicate the picture. A growing number of scientists and beekeepers are seeing the phenomenon through their own lenses, based on funding and treatment possibilities, as they search for answers.

UNEXPLAINED BEE KILLS

There are reports of unexplained bee kills practically every beekeeping season. Symptoms are often diffuse and can be ephemeral. Bees with uncoupled wings (K-wings) and distended abdomens may crawl out of colonies or die with their heads buried in comb cells. Signs may be reminiscent of starvation even when the colony has a good deal of stored honey and pollen. Many times, all the beekeeper can point to is a catastrophic decline in worker population.

These losses are some of the most difficult to grapple with for beekeepers and researchers alike. And similar phenomena fill the world apicultural literature, variously called "Disappearing Disease," "Autumn Collapse," "May Disease," and "Spring Dwindling." The current label is Colony Collapse Disorder or CCD. Unfortunately, these terms are only descriptions of symptoms and do not adequately address the root of the problem. The search for solutions must come, as for human illness, from a detailed history of the patient and the situation.

If you suffer large-scale bee loss and can find little reason for it, carefully document the symptoms to be able to clearly communicate them to others. Include in the written description as much detail as possible about management decisions and environmental conditions that affect the situation. It is important to write the information down; verbal description leads to fuzzy thinking and cannot be referred to by others. Only by analyzing recorded observations will researchers and beekeepers alike have a place to begin the difficult search for solutions to unexplained bee loss.

Wax Moth

The greater wax moth (*Galleria mellonella*) is the honey bee's garbage collector. Its role in nature is to destroy abandoned nests, in the process eliminating possible sources of infestation that might affect healthy nearby colonies. A strong colony of bees will not have wax moth problems, because worker bees are extremely efficient in searching out and eliminating the larvae. The bees cannot kill these larvae, but they will eject them from the hive.

Beekeepers often attribute colony decline directly to wax moth, but this is not usually the case. Rather, a colony weakened by any stress factor will attract adult moths. These work to gain a foothold by laying eggs inside and sometimes nearby a bee colony, which hatch into voracious larvae. If there is not enough worker population to remove the offending larvae, they complete their development and begin reproducing.

The biggest problem beekeepers have with wax moths is not inside colonies, but in stored supers, which are usually infested with eggs that are difficult to see. The larvae from these eggs obtain nutrients from honey, cast-off pupal skins, pollen, and other impurities found in old comb to complete their development, but not from the beeswax itself. Consequently, older combs that contained brood are more likely to be infested than ones only containing honey or frames containing beeswax foundation. Ruined comb results when larvae produce a mass of web, called galleries. Wax moths are an

Adult wax moths are on the left; larvae are on the right.

especially serious problem in tropical and subtropical climates, where year-round high temperatures favor their rapid development. Found anywhere bees might be kept, wax moths are so easy to rear that the larvae are used for fish bait and to feed pet lizards or amphibians.

Symptoms. A heavy infestation of wax moth larvae produces webs and builds large numbers of cocoons for pupation. These tough larvae often damage the wooden frames and hive bodies. In extreme cases, very little structure is left in either comb or frame.

Treatment. Clean up lightly infested combs by introducing them into strong bee colonies. You can renovate heavily infested hives by removing the damaged comb and replacing it with new foundation, if the wooden frame is not too damaged by gouges where pupal cases are attached.

The best defense against wax moths is to keep the colony in vigorous good health through good management. There is continuing interest in biological control using *Bacillus thuringiensis* (Bt), a biological pesticide that kills moth larvae after they ingest it. This is applied as a liquid on combs, but is a practice that many see as too labor intensive.

Control of wax moths in stored combs is possible through use of chemicals (fumigation) and nonchemical methods such as freezing the comb. In many areas, stored combs freely exposed to light and air suffer much less damage than those kept in supers stored in dark, heated buildings.

Small Hive Beetle

Movement of biological material around the globe is becoming epidemic. Examples of organisms introduced by humans are legion, including plants, mosquitoes, beetles, moths, and mussels. *Aethina tumida*, the small hive beetle from South Africa, is just one of the latest examples. Besides being recently introduced into the Americas, it is now well established in Australia.

How the small hive beetle made the long trip from its African homeland to the United States remains unknown. The first sign of its presence was a large number of larvae (worms) appearing in honey-extracting rooms near Fort Pierce on the east coast of Florida in 1998.

Small hive beetle larvae in honey are a recipe for disaster. They change the honey chemically and it begins to quickly ferment; "slimed" combs are then no longer attractive to bees. The honey for all practical purposes is ruined and cannot be recovered. If you rinse the combs with water to remove fermented honey, however, bees will often accept them for refilling.

Treatment. The best time to target the larvae for treatment is when they leave the comb in search of soil in which to pupate. Because they are attracted to light, small hive beetle larvae can easily be drawn to bright areas illuminated by electric light bulbs in a honey house and then shoveled into water, where they quickly drown. Some beekeepers have had success treating the soil around colonies with chemicals. Various traps also have been developed to catch adults, and this form of control appears to have a good future.

The adult beetle is red just after pupation and then turns black. It is fairly uniform in color, moves rapidly across the comb, and is extremely difficult to pick up by hand because it is covered with fine, slippery, hairlike spines.

It is visible to the naked eye and about a third the size of a worker honey bee.

Published references and communications from South African beekeepers and regulatory authorities indicate that beetle infestations in that country are not common and are generally the sign of sloppy beekeeping practices. This is because, as with the wax moth, the beetle is a scavenger and must get a foothold in a colony before it can reach population numbers sufficient to undermine a beehive.

At this time, it is difficult to say how aggressively the beetles are causing colony mortality in the United States. It will take a few years to determine how this nonnative organism will fit into its new American environment. Notably, some experienced beekeepers have seen strong colonies quickly collapse, which they blame directly on the beetle. Adult beetles are strong fliers and are attracted to colonies that are under stress or undersized (nucs), unable to defend themselves from infestation.

LOOK-ALIKE LARVAE

The larvae of the small hive beetle superficially resemble those of the wax moth (*Galleria mellonella*) and were considered the same at first. To complicate things, both wax moth and beetle larvae can occur together in the same bee colony.

Beetles and moths undergo development similar to bees, beginning with an egg, which hatches into a feeding larva that completes its development during a resting stage (pupa), and finally emerges as an adult. On closer inspection, however, it becomes clear that wax moths and small hive beetles are two different creatures.

Wax moth larvae have many small, uniform prolegs behind their real legs, while small hive beetle larvae possess three pairs of larger, more pronounced, real legs on the thorax near the head. Another difference is that the beetle larvae do not grow as large as wax moth larvae before pupation. They also do not spin a cocoon in the hive but must complete their development in the soil.

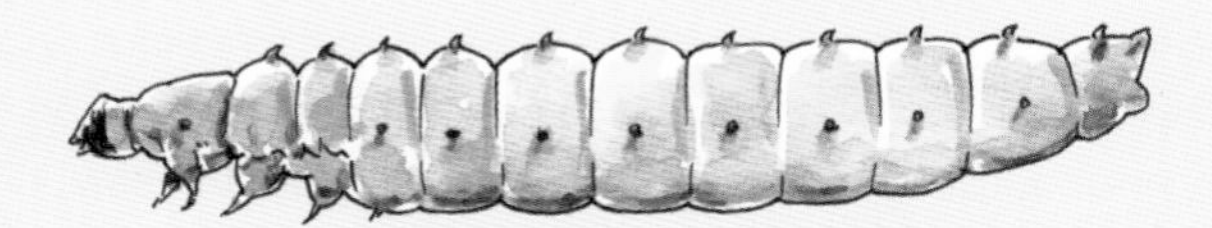

larva, hive beetle

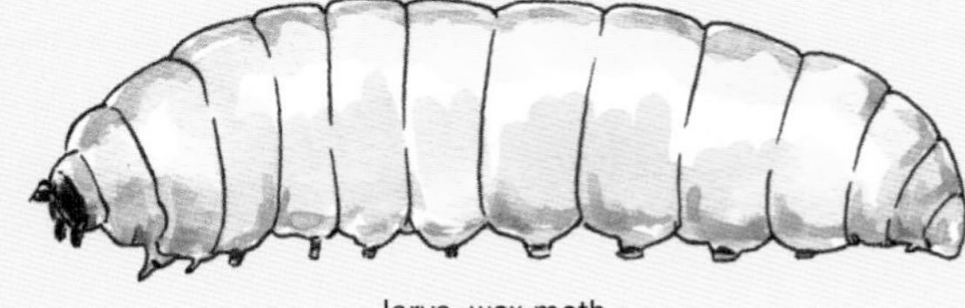

larva, wax moth

The growth stage of Lepidoptera (butterflies and moths) has uniform prolegs (right), whereas the larva of Coleoptera (beetles) has six pronounced true legs on its thorax, or anterior part (above).

The small hive beetle has quickly changed the habits of beekeepers in the honey house. No longer is it possible to leave wet and filled supers for long before and during the extraction process. The advice now is to quickly extract honey and return the wet supers to bee colonies.

Black Bears

The black bear (*Ursus americanus*) can become a big beekeeping pest wherever it is established. Although the bear's appetite for honey is celebrated in popular culture, brood is what it really likes most as a protein source.

Research reveals that bears do not search out beehives but are opportunistic feeders. When they come across unprotected colonies, the bees are relatively defenseless. Bears get stung a lot, but that's not much of a deterrent. They may carry colonies out of the apiary and feast on them after the bees become disoriented and fly back to their original location.

Bears and beekeepers don't mix, and beekeepers do sometimes kill bears if they discover them molesting bee colonies. The damage bears cause to wooden beehives is legendary, and they can make quick work of a carefully groomed apiary. That said, bears are often protected by law, and a beekeeper with a gun risks being fined or arrested for shooting one.

The best strategy to keep bears from contacting hives is by erecting electric fences. They serve as an excellent deterrent, but must be in place before the bear contacts any hives. Once bears have had a taste of bee larvae and/or honey, electric fences are no longer adequate protection. Many ideas and designs have been published over the years, and most are effective.

Report problem bears to wildlife officials. In some regions, bears are trapped and traumatized (a tooth is pulled) before being released in an effort to encourage the bear not to return. Alternatively, officials transport them to more remote locations in the hope that they will not be able to find their way back to the affected beeyard.

BEARS AND BEES

Bears are a heady topic at many beekeeper meetings. Although their images are used to promote honey, practically any mention of these furry mammals is guaranteed to get strong reactions. This is understandable when one is confronted by the devastation a single bear can cause in a beeyard. Mention of eliminating these animals, however, is anathema to those who are interested in conserving this wildlife resource.

The emotional content of the topic is heightened because so little verifiable information is available either on the amount of damage bears cause or the physical number of these animals in any given area. The bear-beekeeper conflict affects at least 39 of 50 states and most Canadian provinces. The means to control black bear damage in descending order of effectiveness are electric fences, platforms to elevate hives, and hunting.

The writing is on the wall: black bear populations are here to stay. In the long run, paradoxically, a concerted effort to save the black bear will also benefit the beekeeper by preserving wild habitat essential to many nectar- and pollen-producing plants.

For More Information

If you find a disease or parasite unknown to area beekeepers, consider sending affected bees to a laboratory specializing in honey bee pathology. Traditionally, the Beltsville, Maryland, Bee Laboratory has serviced the national beekeeping community; some private labs, however, have begun analyzing honey bees (see Resources).

COMPARATIVE SYMPTOMS OF HONEY BEE BROOD DISEASES

Symptom	American foulbrood	European foulbrood	Sacbrood	Chalkbrood
Appearance of brood comb	Sealed brood. Discolored, sunken, or punctured cappings.	Unsealed brood. Some sealed brood in advanced cases with discolored, sunken, or punctured cappings.	Sealed brood. Scattered cells with punctured cappings.	Sealed and unsealed brood.
Age of dead brood	**Usually older sealed larvae or young pupae.***	**Usually young unsealed larvae; occasionally older sealed larvae.** Typically in coiled stage.	Usually older sealed larvae; occasionally young unsealed larvae. Upright in cells.	Usually older larvae. Upright in cells.
Color of dead brood	Dull white, becoming light brown, coffee brown to dark brown, or almost black.	Dull white, becoming yellowish white to brown, dark brown, or almost black.	Grayish or straw-colored, becoming brown, grayish black, or black. Head end darker.	**Chalk white.** Sometimes mottled with black spots.
Consistency of dead brood	**Soft, becoming sticky to ropy.**	**Granular.** Watery; rarely sticky or ropy.	**Watery and granular; tough skin forms a sac.**	Watery to pastelike.
Odor of dead brood	Slight to pronounced odor.	Slightly sour to penetratingly sour.	None to slightly sour.	Slight, non-objectionable.
Scale characteristics	Uniformly lies flat on lower side of cell. Adheres tightly to cell wall. Fine, threadlike tongue of dead pupae may be present. Head lies flat. Brittle. Black.	Usually twisted in cell. Does not adhere tightly to cell wall. Rubbery. Black.	Head prominently curled toward center of cell. Does not adhere tightly to cell wall. Rough texture. Brittle. Black.	**Does not adhere to cell wall.** Brittle. Chalky white, mottled, or even black.

* **Bold** type indicates most significant symptoms.

Source: Shimanuki, H. and D. A. Knox, *Diagnosis of Honey Bee Diseases, Agriculture Handbook Number 690, Agricultural Research Service, United States Department of Agriculture*

Dadant

11

ADDITIONAL STRATEGIES

This book is written for the novice beekeeper and presents "standard" beekeeping practice. As noted previously, the Reverend L. L. Langstroth recognized the significance of the bee space and developed the movable-frame hive that bears his name. The resulting standard beekeeping approach has been around for more than 150 years and has served the world beekeeping community well. Now it is the norm, and its components are easily purchased, interchangeable, and suited to large-scale commercial honey and pollination operations that move beehives long distances.

Unfortunately, beekeepers face a much more complex environment now than at any time in the history of the craft. In response, they are employing novel technologies and cutting-edge information from a variety of fields to ensure that apiculture prospers in the future. For example, researchers are studying the honey bee's biology in more depth than previously possible, thanks to the recent decoding of the insect's DNA.

Also invigorating the beekeeping environment is a philosophical shift that includes the examination of other beekeeping techniques. On the table is everything from beehive design to management practices that support the honey bee's innate defense mechanisms against diseases and pests. We will explore some of these ideas in this chapter.

Beehive Design

The standard Langstroth-type beehive has a vertical structure, with a basal brood chamber topped with other boxes called "supers." This has many benefits for honey bees, including better population buildup, less crowding and swarming, easier comb building, and minimal disturbance when surplus honey is removed.

Increasingly, however, small-scale and other beekeepers are now exploring different hive designs. Many of these, along with changes in management style, are often considered to be more biologically in tune with a honey bee colony than the Langstroth archetype. We'll describe some of these in this section, because they cannot be ignored in the current beekeeping environment.

A beekeeper brushes a frameless comb removed from a top bar hive.

top bar hive

The Top Bar Hive: a Kinder, Gentler Beekeeping

Dr. Wyatt Mangum, an international bee biologist, is a pioneer in top bar beekeeping and has convinced many committed Langstroth-standard hive users that this approach has a place in their operation. The model appeals to the experimenter and tinkerer often living inside every beekeeper. The **top bar hive** (**TBH**) is inexpensive and can be made out of scrap lumber — perfect for an energetic and motivated new beekeeper.

The top bar is also a flexible system. Dr. Mangum can reduce his larger hives' weight with a buck saw — he simply cuts several bars off the end, and then converts the cut end to a queen-rearing nucleus. This is perhaps the ultimate in hive flexibility. Imagine taking a standard Langstroth ten-frame hive and converting it to an eight-frame model with a two-frame nucleus left over in the bargain.

TBH beekeeping is easy on both bees and beekeeper, according to Dr. Mangum. There are a number of reasons for this.

- The brood is generally placed toward the front-entrance end of the hive and the honey is located in the rear. Examining the brood or taking off honey is, therefore, less stressful on the insects because one doesn't have to dismantle the whole colony.
- The top bars butt against each other. Because of this they double as a cover, reducing material requirements and conserving weight. An outer cover of tin or cardboard is necessary, however, to protect the colony from moisture.
- Only the part of the hive being worked is exposed during manipulation, which reduces overall defensiveness.
- Finally, all Dr. Mangum's hives are mounted on stands at waist level, keeping him from having to continually bend over.

This design does not use entire frames — the bees are provided only with the top bars (thus the name). The horizontal hive body has angled walls, which honey bees do not attach to the resultant combs. "Frameless" beekeeping

Examining a frame from a horizontal or long hive

means that much less material and knowledge are needed to construct a beehive. This design is particularly suitable for use in the developing world, where resources are slim, such as Latin America and Africa. The top bar hive is often called the "Kenya Top Bar Hive," after the African country where it was first extensively used.

It is notable that Dr. Mangum does not employ honey supers in his operation. The possibility exists, however, to use them as part of a hybrid system, made up of a horizontal style brood chamber coupled with standard honey supers stacked vertically. Many long hives in fact use this design, where both the brood chamber and supers have conventional frames.

Kenyans developed the method of top bar beekeeping. Tree of Life Beekeeping, a nongovernmental organization, teaches this method of beekeeping all over East Africa.

Horizontal or Long Hives

Hives with full frames can also be managed in a horizontal way. Many of the advantages seen in the top bar hive described by Dr. Mangum are also realized using this design. Not often seen in the U.S., horizontal, sometimes called "long," hives are much more common in European beekeeping.

A pioneer in the horizontal style is Georges de Layens, who wrote a classic book on the subject that has been translated from the French. His system consists of a single hive of twenty frames, each frame being 13 inches long and 16 inches deep. This is a clear departure from the standard Langstroth ten-frame hive, where each frame is 19 inches by 9⁹⁄₁₆ inches. Specifically, the depth of the Layens frame is almost twice that of the standard design, providing benefits to honey bees over vertical systems that may not at first be obvious.

A major plus is that the comb is not divided over the extent of its depth. Vertical systems create one or more artificial barriers that honey bees have to cross as they navigate the hive's complex of brood chamber and associated complex of stacked boxes (supers). These gaps or barriers are not found in feral or wild honey bee nests, which often have extremely long, deep combs built into narrow tree trunks. Among other advantages, therefore, the Layens design helps honey bees overwinter successfully, because the cluster is not forced to move across spaces between boxes.

The Layens hive and associated management system may also appeal to both part-time and novice beekeepers who aren't yet experienced in managing honey bees using a vertical model. Taken together, the design of the horizontal system found in both the top bar model described by Dr. Mangum and the Layens hive can result in a much more traditional hands-off approach.

De Layens's conclusion concerning successful beekeeping is just as apt today as it was when he published his treatise: "We cannot improve beekeeping by going farther and farther away from the bees' natural tendencies. Instead, pick the hive model that is best matched to your locale, populate it with local bees, and the results will speak for themselves."

Other Archetypes

Other beehive systems exist, ranging from the traditional honey bee management in logs (gums) and skeps to the more esoteric. Among these are the Warré and Sun hives. Each has a devoted, often passionate, following.

The Warré hive resembles a Langstroth hive in its vertically stacked boxes, but instead of full frames the combs are supported by top bars. To give the colony more room as it grows, the beekeeper inserts a new box under the bottom box, rather than placing it on top.

The Sun Hive (*Weissenseifener haengekorb*) turns traditional beekeeping on its head and appears to be a throwback to "fixed comb" systems of the past. Honey production is not the aim. Rather, conserving the honey bee as a species is the goal. It is a project of The Natural Beekeeping Trust and incorporates ideas from two fairly recent books, *Honeybee Democracy* by Dr. Thomas Seeley and *The Buzz About Bees* by Jürgen Tautz.

Inspecting a frame from a Warré hive

Alternative Management Practices

So-called biodynamic beekeeping is considered apicentric in that it seeks to minimize environmental stress factors, often exacerbated by "industrial" practices. This allows honey bees to exhibit their true nature. The following guidelines are published by the Rudolf Steiner College's beekeeping curriculum (rudolfsteinercollege.edu/beekeeping) in an effort to encourage more environmentally friendly beekeeping:

- Natural combs are used, rather than foundation.
- Swarming is recognized as the natural form of colony reproduction.
- Clipping of queen's wings is prohibited.
- Regular and systematic queen replacement is prohibited.
- Pollen substitutes are prohibited.
- Beehives must be made of all-natural materials, such as wood, straw, or clay.
- Artificial insemination is not used. Instead queens are allowed to fly free to mate.
- Grafting of larvae to produce queens is prohibited.
- No pesticides or antibiotics are allowed, although the use of natural organic acids such as formic and oxalic acid may be used for mite control.
- Honey may be transported in containers made of artificial materials but must be decanted into containers of glass or metal for retail sale.

Current beekeeping management systems incorporate many of the practices adopted from conventional agricultural techniques. Especially controversial is the routine use of antibiotics and pesticides. Unfortunately, these chemicals often have unintended consequences, including bypassing, and in some cases inactivating, many of the formidable defense mechanisms naturally found in honey bees (see pages 159–161). Large-scale commercial pollination also stresses honey bee colonies in several ways that do not contribute positively to honey bee health.

Apicentric beekeeping emphasizes the power of natural selection, encouraging the development of what are called "local" honey bees as noted previously by Georges de Layens. Specific mechanisms have already been discovered in some stock considered tolerant to diseases and pests (see pages 179–180). Forced natural selection as found in countries that do not treat honey bees with chemicals, including South Africa, Brazil, and Cuba, has also resulted in what many call "survivor" honey bees. Research by Dr. Thomas Seeley, author of *Honeybee Democracy*, continues to show populations of survivor bees spontaneously developing in the forests near Cornell University. Similar reports indicate that this might be occurring elsewhere.

Entomologist John Kefuss developed the "Bond test" as a strategy to select genetics for resistance to varroa mites and other maladies in honey bees. The original "live and let die" method, which has garnered a lot of attention from beekeepers and researchers alike, withheld chemical treatment and removed from the breeding population all honey bees that were diseased, weak, or previously chemically treated yet still struggling. Kefuss currently promotes a "modified Bond" technique in which colonies with high mite levels are in fact treated, to prevent spreading mites to other hives.

This brings us to the current conversation in the beekeeping community about "survivor" or "resistant" stock, often conflated with "treatment-free" beekeeping. As with many issues in the modern world, managing honey bees to achieve either or both goals is complex. A beekeeper who embarks on a treatment-free protocol must also use resistant or survivor stock as an integral part of the strategy. And there should be a plan in hand, similar to Dr. Kefuss's "modified Bond" treatment above, to treat or euthanize any affected hives before their collapse can impact others nearby.

Toward a Honey Bee–Friendly Future

The honey bee is an essential contributor to the health and well-being of our planet. Every day, farmers, scientists, and average citizens grow more committed to seeing this social insect thrive. No pledge, however, is more important in the current environment than to become a beekeeper.

The goal of this book is to provide the novice apiculturist with a comprehensive guide to working alongside honey bees thoughtfully and with clear purpose. In the process, we hope to communicate some of the formidable challenges, along with the thrills, that humanity has long experienced in its dedication to helping one of nature's most complex creatures flourish and prosper.

GLOSSARY

abdomen. The third section of the honey bee body, containing the intestine (ventriculus), reproductive and excretory systems, and sting apparatus.

absconding. When a colony leaves its present location in search of another, usually brought on by adverse environmental conditions such as lack of food and water, excessive harassment, or presence of disease.

Acarapis woodi. Scientific name of the honey bee tracheal mite.

Africanized honey bee. Common name of the tropical hybrid honey bee based on the African honey bee ecotype, *Apis mellifera scutellata*, introduced to Brazil in the 1950s.

American foulbrood. A notoriously lethal bacterial disease of the brood.

apiculture. Another word for beekeeping; culturing the honey bee (genus *Apis*).

Apis mellifera. Scientific name of the Western honey bee.

beebread. Stored food in the colony based on fermented pollen mixed with small amounts of other nutrients.

bee space. The space conserved by the honey bee to provide passageway through the comb. About 5⁄16 inch wide, this space is the basis for development of the movable-frame hive on which modern beekeeping is based.

brood. Collectively, the immature stages of the honey bee, including eggs, larvae, and pupae.

brood chamber. A section of the hive used for brood rearing.

capped brood. The portion of the brood that has been covered with wax and houses the pupae.

chalkbrood. A fungal infection of the honey bee brood caused by *Ascosphaera apis*.

comb. Collectively, the wax cells of a nest, which are constructed back to back into a solid slab and usually surrounded by a wooden frame.

crop. A modified portion of the esophagus that holds nectar. It is the cargo hold of the honey bee.

emergence. The point in a honey bee's life when it has completed its metamorphosis and cuts its way out of the cell to join the other adult inhabitants in a colony.

European foulbrood. A bacterial disease of the brood caused by *Melissococcus pluton*.

foundation. Thin sheets of beeswax (sometimes plastic) embossed with hexagonal cells. It is the template used by honey bees to construct their cells, which become the comb.

honey. Sweet material collected first as nectar and then modified for storage and consumption by the colony. The surplus is then harvested by the beekeeper.

honey flow. When large quantities of nectar are available from plants to convert into honey.

hypopharyngeal glands. Glands in the head of worker honey bees that produce royal jelly or bee milk. Sometimes called brood food glands.

nectar. Sweet sugar solution secreted by flowers that is the raw material for honey.

nosema. A fungus that infects the midgut or ventriculus, found in adult bees. It is caused by two variants, *Nosema apis* and *Nosema ceranae*.

nucleus hive (nuc). A nucleus colony is a starter unit (three to five frames of bees) that is usually installed into a standard 10-frame hive, and becomes the foundation for a new full-sized colony. It is often called a "nuc."

open brood. Brood that has yet to be capped with wax, containing eggs and larvae.

package bees. A screened container containing a specific number of adult honey bees sold to start new colonies. It usually comes with a caged queen and a can of sugar syrup.

parthenogenesis (virgin birth). The development of an embryo from an unfertilized egg, as occurs with the drone.

pheromone. A chemical substance secreted by an organism that elicits specific behavior in others of its species. Queen pheromones in honey bees keep worker bees from laying eggs.

pollen. Dust-like grains produced in the anthers of flowers as the male element of their reproductive system. These are collected and stored by honey bees as the protein part of their diet.

propolis. Resins collected by honey bees from certain trees used for filling cracks and crevices in the comb. Used as a comb reinforcement and to cover and isolate objectionable materials bees cannot physically remove from a hive.

queen excluder. A device used to restrict movement of queens in a hive. It also restricts drones.

queenright. Describes a colony that has an active, healthy queen.

royal jelly. A nutritious food (sometimes called bee milk) produced by worker bees, which is fed to larvae. It is given in prodigious quantities to larvae that will develop into queens.

sacbrood. A viral infection of honey bee brood.

small hive beetle (*Aethina tumida*). A pest that destroys honey, pollen, and comb and may force bees to leave their hive.

spermatheca. An organ in queens containing sperm she received from mating with drones.

swarming. The reproductive act in which a portion of a honey bee colony (a swarm), including the queen, leaves its home to establish a new one elsewhere. The colony left behind (parent) produces a new queen and continues on. Different than absconding, where the entire colony simply deserts its nest.

thorax. The second or middle part of an insect's body, between head and abdomen, which is dedicated to locomotion and heat production. The thorax contains the wing and leg muscles.

tracheal mite (*Acarapis woodi*). A mite that inhabits the adult honey bee's respiratory system, feeding on its blood, clogging the breathing tubes, and eventually killing the insect.

tracheal tubes. Breathing tubes, sometimes called tracheoles, in insects that are used to conduct air to all cells in the body.

varroa mite. The exotic Asian honey bee mite, previously called *Varroa jacobsoni*, but now known as *Varroa destructor*.

SAMPLE POLLINATION CONTRACT

The following is a suggested pollination agreement adapted from that found in Agriculture Handbook 496, *Insect Pollination of Cultivated Crop Plants*, by S. E. McGregor, USDA, 1976.

This agreement is made ________________________, 20___, between ________________________ __________, (grower's name), hereinafter called the grower, and ______________________________ (beekeeper's name), hereinafter called the beekeeper.

1. TERM OF AGREEMENT. The term of this agreement shall be for the ______________ growing season.

2. RESPONSIBILITIES OF THE BEEKEEPER

A. The beekeeper shall supply the grower with ________ hives (colonies) of bees to be delivered to the ________________________________ (cucumber, watermelon field, etc.) as follows:
(Fill in the appropriate line or lines and cross out those that do not apply.)
Approximate date: ___________________ or ______ days after written notice from the grower.
Time in relation to amount of crop bloom: ______________________________________
____

Description of location(s): ____
____
____
(If additional space is needed, attach separate sheet dated and signed by both parties.)

The beekeeper shall locate said bees in accordance with directions of the grower, or, if none are given, according to his judgment in order to provide maximum pollination coverage.
B. The beekeeper agrees to provide colonies of the following minimum standards:
A laying queen with the following:
______________ frames with brood with bees to cover.
______________ pounds (_______ kg) of honey stores or other food.
______________-story hives.
The grower shall be entitled to inspect, or cause to be inspected, each colony of bees after giving reasonable notice to the beekeeper of this intent.
C. The beekeeper agrees to maintain the bees in proper pollinating conditions by judicious inspection with supering or honey removal as needed.
D. The beekeeper agrees to leave the bees on the crop until:
(Fill in the appropriate line or lines and cross out those that do not apply.)
Approximate date: _____________________ or _________ days after written notice from the grower.
Time in relation to amount of crop bloom: ______________________________________
____

Other: ____

3. RESPONSIBILITIES OF THE GROWER

A. The grower agrees to provide a suitable place to locate the hives. The site must be accessible by a truck and other vehicles used in handling and servicing the colonies. The grower shall allow the beekeeper entry on the premises whenever necessary to service the bees, and the grower assumes full responsibility for all loss and damage to his fields or crops resulting from the use of trucks or other vehicles in handling and servicing such bees.

B. The grower agrees not to apply highly toxic pesticides to the crop while the bees are being used as pollinators immediately prior to their movement if the residue would endanger the colonies. The following pesticide materials, other agricultural chemicals, and methods of application are mutually agreed to be suitable while the bees are on the crop: ______________________________

The grower agrees to notify the beekeeper if hazardous materials not listed are to be used. The cost of moving the bees away from and back to the crop to prevent damage from highly toxic materials shall be borne by the grower.

C. The grower agrees to pay for __________ colonies of bees at the rate of $______________ per colony. Payment shall be made to the beekeeper as follows:

$______________ per colony on delivery and the balance on or before ______________ of said year.

Additional moves or settings shall require $______________ per hive per move.

D. The grower agrees to provide adequate watering facilities for the bees if none are available within one-half mile (0.8 km) of each colony used in pollinating the crop.

4. PERFORMANCE. It is understood and agreed that either party to this agreement shall be excused from the performance hereof in the event that, before delivery of the colonies, such performance is prevented by causes beyond the control of such party.

5. ARBITRATION. If any controversy shall arise hereunder between the parties hereto, such controversy shall be settled by arbitration. Each party within 10 days shall appoint one arbitrator, and the so-named shall select a third, and the decision by any two such arbitrators shall be binding upon the parties hereto. The cost of such arbitration shall be divided equally between the parties.

6. ASSIGNMENT OR TRANSFER. This agreement is not assignable or transferable by either party, except that the terms hereof shall be binding upon a successor by operation of law to the interest of either party.

IN WITNESS WHEREOF, the parties hereto have executed this agreement the day and year above.

Grower ____

By ____

(address) ____

Beekeeper ____

By ____

(address) ____

MODEL BEEKEEPING ORDINANCE

If your town or city is considering beekeeping ordinances, the following model may be acceptable to both local government and beekeepers.

Section 1. Location of Beehives and Other Enclosures

It shall be unlawful for any person to locate, construct, reconstruct, alter, maintain, or use on any lot or parcel of land within the corporate limits, any hives or other enclosures for the purpose of keeping any bees or other such insects unless every part of such hive or enclosure is located at least seventy-five (75) feet from a dwelling located on the adjoining property.

Section 2. Number of Hives (Colonies of Bees) Regulated

On lot sizes of 15,000 square feet or less, no more than four hives (colonies of bees) will be permitted. The hives shall be no closer than 15 feet from any property line. On lots larger than 15,000 square feet, additional hives will be permitted on the basis for one (1) hive for each 5,000 square feet in excess of 15,000 square feet.

Section 3. Type of Bees

This ordinance shall pertain only to honey bees maintained in movable-frame hives, and it does not authorize the presence of hives with non-movable frames or feral honey bee colonies (honey bees in trees, sides of houses, etc.).

Section 4. Restrictions on Manipulating Bees

The hives (colonies) of bees may not be manipulated between the hours of sunset and sunrise unless the hives are being moved to or from another location.

Section 5. Penalty

The violation of any provision of this ordinance shall constitute a misdemeanor punishable upon conviction by a fine not exceeding fifty ($50) dollars, or imprisonment not exceeding thirty (30) days; provided that each day that a violation exists or continues to exist shall constitute a separate offense.

Section 6. Effective Date

This ordinance shall be effective from and after the ____________ of _______________, 20___.

A Sampling of U.S. Beekeeping Supply Houses

Betterbee
Greenwich, New York
800-632-3379
www.betterbee.com

Blue Sky Bee Supply Ltd.
Ravenna, Ohio
877-529-9233
www.blueskybeesupply.com

Brushy Mountain Bee Farm
Moravian Falls, North Carolina
800-233-7929
www.brushymountainbeefarm.com

Cook & Beals
Loup City, Nebraska
308-745-0154
www.cooknbeals.com

Cowen Manufacturing Co.
Parowan, Utah
800-257-2894
www.cowenmfg.com

Dadant and Sons
Hamilton, Illinois
888-922-1293
www.dadant.com

Deb's Bee Supply
Jacksonville, Florida
877-703-3327
www.deb-bee.com

Draper's Super Bee Apiaries
Millerton, Pennsylvania
800-233-4273
www.draperbee.com

Kelley Beekeeping
Clarkson, Kentucky
800-233-2899
www.kelleybees.com

Mann Lake Ltd.
Hackensack, Minnesota
800-880-7694
www.mannlakeltd.com

Maxant Industries
Ayer, Massachusetts
978-772-0576
www.maxantindustries.com

Miller Bee Supply
North Wilkesboro, North Carolina
336-670-2249
www.millerbeesupply.com

Pierco Beekeeping Equipment
Riverside, California
800-233-2662
www.pierco.net

Rossman Apiaries
Moultrie, Georgia
800-333-7677
www.gabees.com

Ross Rounds Inc.
Canandaigua, New York
518-370-4989
www.rossrounds.com

Western Bee Supplies
Polson, Montana
406-883-2918
www.westernbee.com

Sources of Beekeeping Information

Encyclopedias

Graham, Joe M., and Dadant & Sons, eds. *The Hive and the Honey Bee*, rev. ed. Dadant & Sons, 2015.

Shimanuki, Hachiro, Kim Flottum, and Ann Harman, eds. *ABC & XYZ of Bee Culture*, 41st ed. Bee Culture Publications, 2007.

Periodicals

American Bee Journal

Dadant & Sons, Inc.
217-847-3324
www.americanbeejournal.com

Bee Culture

A. I. Root Company
800-289-7668
www.beeculture.com
Includes a virtual, continuously updated version of *Insect Pollination of Cultivated Crop Plants*, by S. E. McGregor

Recommended Books

Caron, Dewey M. *Honey Bee Biology and Beekeeping*, 2nd ed. Wicwas Press, 2001.

Connor, L. J. *Increase Essentials*. Wicwas Press, 2006.

Delaplane, Keith, and Daniel Mayer. *Crop Pollination by Bees*. Washington State University, 2000.

Flottum, Kim. *The Backyard Beekeeper: An Absolute Beginner's Guide to Keeping Bees in Your Yard and Garden*. Quarry Books, 2010.

———. *The Backyard Beekeeper's Honey Handbook: A Guide to Creating, Harvesting and Cooking With Natural Honeys*. Quarry Books, 2009.

Guzman-Novoa, E. *Elemental Genetics and Breeding for the Honeybee*. Ontario Beekeepers' Association, 2007.

Hoopingarner, R. *The Hive and the Honey Bee Revisited*. Bee+ Books, 2006.

Horn, Tammy. *Bees in America: How the Honey Bee Shaped a Nation*. University Press of Kentucky, 2005.

Hubbell, Sue. *A Book of Bees*. Houghton Mifflin, 1988.

Johansen, Carl A., and Daniel F. Mayer. *Pollinator Protection: A Bee and Pesticide Handbook*. Wicwas Press, 1990.

Laidlaw, Harry H., and Robert E. Page. *Queen Rearing and Bee Breeding*. Wicwas Press, 1998.

Loring, Murray. *Bees and the Law*. Dadant Publications, 1981.

Lovell, Harvey. *Honey Plants of North America*. A. I. Root Co., 1926.

Matheson, Andrew. *Living with Varroa*. International Bee Research Association, London, 1993.

Morse, Roger A. *Raising Queen Honey Bees*, 2nd ed. Wicwas Press, 1993.

———. *Making Mead (Honey Wine)*. Wicwas Press, 1992.

Sammataro, Diana, and Alphonse Avitabile. *The Beekeeper's Handbook*, 3rd ed. Comstock Publishing, 1998.

Seeley, Thomas D. *Honeybee Democracy*. Princeton University Press, 2010.

Spivak, M., D. J. C. Fletcher, and M. D. Breed, eds. *The "African" Honeybee*. Westview Press, 1991.

Taylor, Richard. *The How-To-Do-It Book of Beekeeping*, 5th ed. Linden Books, 1998.

Tautz, Jurgen. *The Buzz about Bees: Biology of a Superorganism*. Springer, 2009.

von Frisch, Karl. *Bees: Their Vision, Chemical Senses and Language*, rev. ed. Cornell University Press, 1971.

Winston, Mark L. *The Biology of the Honeybee*, 5th ed. Harvard University Press, 1995.

Woodward, David. *Queen Bee: Biology, Rearing and Breeding*. Telford Rural Polytechnic, 2007.

A Few Blogs of Interest

www.scientificbeekeeping.com
http://peacebeefarm.blogspot.com
www.pollinatethis.org/beeblog
http://mellifera.blogspot.com
www.hive-mind.com/backyard-beekeeping
http://beekeeperlinda.blogspot.com
http://basicbeekeeping.blogspot.com
http://biobees.blogspot.com
www.honeybeeworld.com/diary
www.badbeekeepingblog.com
www.beekeep.info/apis-newsletter

National Beekeeping Organizations

American Beekeeping Federation

404-760-2875
www.abfnet.org

American Honey Producers Association

281-900-9740
www.ahpanet.com/

Works Cited in Text

American Association of Professional Apiculturists (AAPA). Position Statement on the Health of the U.S. Honey Bee Industry. http://aapa.cyberbee.net/wp-content/uploads/2013/06/Final_AAPA_POSITION_STATEMENT_COLONY_HEALTH-1-1.pdf.

Banks, David, and Barbara Waterhouse. *An Introduction to Beelining*. Australian Quarantine and Inspection Service. www.bindaree.com.au/hints/hint12_beelining.

Camazine, Scott. "Hymenopteran Stings: Reactions, Mechanisms, and Medical Treatment." *Bulletin of the Entomological Society of America* (Spring 1988): 17–21.

de Layens, Georges. *Keeping Bees in Horizontal Hives: A Complete Guide to Apiculture*. Translated by Mark Pettus. Deep Snow Press, 2017.

Dustmann, Josh. "Natural defense mechanisms of a honey bee colony against diseases and parasites," *American Bee Journal* 133, no. 6 (June 1993): 431–34.

The Honey Bee Health Coalition, guide on varroa control. https://honeybeehealthcoalition.org/wp-content/uploads/2015/08/HBHC-Guide_Varroa-Interactive-PDF.pdf

Mangum, Wyatt A. *Top-Bar Hive Beekeeping: Wisdom & Pleasure Combined*. Signature Book Printing, 2012.

McGregor, S. E. *Insect Pollination of Cultivated Crop Plants*. (USDA, 1976). Out of print; online at www.beeculture.com/content/pollination_handbook/index.cfm.

Sanford, Malcolm T. *Beekeeping without Borders: Apiculture in Italy and France at the Dawn of the European Union*. Northern Bee Books, 2016.

Seeley, Thomas D. *Honeybee Democracy*. Princeton University Press, 2010.

Tautz, Jürgen. *The Buzz About Bees: Biology of a Superorganism*. Springer Science & Business Media, 2008.

Warré, Abbé Émile. *Beekeeping for All*. 12th ed. (1948), translated by Patricia and David Heath. Northern Bee Books, 2010.

Winston, Mark. *The Biology of the Honey Bee*, 5th ed. Harvard University Press, 1995.

ADDITIONAL INTERIOR PHOTOGRAPHY CREDITS

© 13-Smile/iStock.com, 63 bottom right; © Alis Photo/Alamy Stock Photo, 16; © allotment boy 1/Alamy Stock Photo, 63 bottom left; © Arterra Picture Library/Alamy Stock Photo, 194; © bernardbodo/iStock.com, 6–7; © Bess Brownlee/Alamy Stock Photo, 188, 193; © Bildarchiv Hansmann/INTERFOTO/Mary Evans Picture Library, 22; © Bramwell Flora/Alamy Stock Photo, 62 top left; © brandtbolding/iStock.com, 33; © ClarkandCompany/iStock.com, 136; © CLAVIERES Virginie/Getty Images, 54; © David Broadbent/Alamy Stock Photo, 100; © David Wootton/Alamy Stock Photo, 118, 192; © DEA PICTURE LIBRARY/Getty Images, 61 bottom right; © DEA/RANDOM/Getty Images, 60 top right; © Dendron/iStock.com, x; © Design Pics Inc/Alamy Stock Photo, 63 top right; © dszc/iStock.com, 63 top left; © ERIC FEFERBERG/Getty Images, 64; © EuToch/iStock.com, 62 middle left & right; © Flowerphotos/Getty Images, 60 top left; © Gérard LABRIET/Getty Images, 61 bottom left; © gkuchera/iStock.com, 153; James D. Ellis/University of Georgia/USDA APHIS Plant Protection and Quarantine program/Wikimedia Commons, 139; © Jerry Harpur/GAP Photos, 61 top right; © joel zatz/Alamy Stock Photo, 62 top right; © Jul Nichols/iStock.com, 62 bottom right; Migco/Wikimedia Commons, 143; © mirceax/iStock.com, 124; © Orjan F. Ellingvag/Getty Images, 14; © Pat Canova/Alamy Stock Photo, 61 top left; © Paul Starosta/Getty Images, 98; © Podius/iStock.com, 78; © SergeyChayko/iStock.com, 150; © Tetsuya Tanooka/Aflo/Getty Images, 60 bottom left; © Tolimir/iStock.com, 30; © vicm/iStock.com, 62 bottom left; © westhimal/123RF Stock Photo, 60 bottom right; © Wolfgang Kaehler/Getty Images, 138

Other Resources

American Apitherapy Society
www.apitherapy.org
Provides information on various treatments using bee products.

Apis Information Resource Center
www.beekeep.info
The author's website includes an extensive guide to honey bees and beekeeping, "A Treatise on Modern Honey Bee Management," and other resources of interest to the discriminating beekeeper.

Bee Health
Cooperative Extension System
www.extension.org/bee_health
A comprehensive website dedicated to honey bee health.

Bee Informed Partnership (BIP)
www.beeinformed.org
Conducts yearly loss surveys and is developing a sentinel apiary program, which monitors honey bee colony health.

Beesource Beekeeping
www.beesource.com
One of the oldest and most comprehensive sites on the Internet. Contains a variety of forums and other information resources, including some controversial information in its Point of View section.

International Bee Research Association (IBRA)
www.ibra.org.uk
A clearinghouse of beekeeping information

National Honey Board
303-776-2337
www.honey.com
Extensive guide to honey research and marketing efforts.

Natural Beekeeping Trust
http://www.naturalbeekeepingtrust.org/bee-centred-vs-conventional
Provides information on bee-centered beekeeping.

Project Apis m.
www.projectapism.org
Funds and directs research to enhance the health and vitality of honey bee colonies while improving crop production.

Rudolf Steiner College
http://rudolfsteinercollege.edu/beekeeping
Teaches biodynamic beekeeping.

Testing Agencies

Bee Research Laboratory
Agricultural Research Service, USDA
Beltsville, Maryland
301-504-8749
www.ars.usda.gov/northeast-area/beltsville-md/beltsville-agricultural-research-center/bee-research-laboratory

North Carolina State University Queen and Disease Clinic
https://entomology.ces.ncsu.edu/apiculture/queen-disease-clinic
Offers a wide range of analytical tools to quantify queen reproductive quality as well as queen and colony health.

METRIC CONVERSION CHARTS

VOLUME CONVERSION

To convert	to	multiply
teaspoons	milliliters	teaspoons by 4.93
tablespoons	milliliters	tablespoons by 14.79
fluid ounces	milliliters	fluid ounces by 29.57
cups	milliliters	cups by 236.59
cups	liters	cups by 0.24
pints	milliliters	pints by 473.18
pints	liters	pints by 0.473
quarts	milliliters	quarts by 946.36
quarts	liters	quarts by 0.946
gallons	liters	gallons by 3.785

VOLUME EQUIVALENTS

U.S.	Metric
1 teaspoon	5 milliliters
1 tablespoon	15 milliliters
¼ cup	60 milliliters
½ cup	120 milliliters
1 cup	240 milliliters
1¼ cups	300 milliliters
1½ cups	350 milliliters
2 cups	480 milliliters
2½ cups	600 milliliters
3 cups	700 milliliters

TEMPERATURE CONVERSION

To convert Fahrenheit to Celsius, subtract 32 from Fahrenheit temperature, multiply by 5, then divide by 9.

Easy-to-Remember Equivalents	
0°C = 32°F	30°C = 86°F
10°C = 50°F	40°C = 104°F
20°C = 68°F	Every 10°C = 18°F

WEIGHT CONVERSION

To convert	to	multiply
ounces	grams	ounces by 28.35
pounds	grams	pounds by 453.6
pounds	kilograms	pounds by 0.45

WEIGHT EQUIVALENTS

U.S.	Metric
0.035 ounce	1 gram
¼ ounce	7 grams
½ ounce	14 grams
1 ounce	28 grams
1¼ ounces	35 grams
1½ ounces	43 grams
1¾ ounces	50 grams
2½ ounces	70 grams
3½ ounces	100 grams
4 ounces	113 grams
5 ounces	140 grams
8 ounces	227 grams
8¾ ounces	250 grams
10 ounces	284 grams
15 ounces	425 grams
16 ounces (1 pound)	454 grams

LENGTH / AREA CONVERSION

To convert	to	multiply by
inches	centimeters	2.54
feet	meters	0.31
yards	meters	0.91
miles	kilometers	1.61
square feet	square meters	0.09
acres	hectares	0.41

INDEX

Page numbers in *italic* indicate drawings and photographs.
Page numbers in **bold** indicate tables and charts.

Q

R

S